Praise for *A Global Warming*

A uniquely powerful resource [that] presents the science of climate change in a way that will be understandable to readers without scientific training . . . a joy to read.
— *Physics Today*

Rarely has a book laid out all of the facts about climate change on Earth so clearly . . . this work ought to be required reading for anyone who actually cares about Earth as a planet.
— **David Eicher,** Editor, *Astronomy Magazine*

There are many explanations available of how global warming works, but few are as clear, comprehensive, and visually descriptive as in this book. Adults of all political and ideological backgrounds will appreciate the book's mature take on the arguments against climate change.
— *Foreword* **Reviews**

With awe-inspiring clarity and solid research, the author presents current information about global warming [science, consequences, and solutions]. This is an extraordinary example of perceptive thinking that encourages readers to weigh the evidence for themselves.
— *School Library Journal*

Bennett's language is simple, direct and clear . . . if you know anyone who is unsure about climate change, or wants a short, "basic-level" introduction to the subject, this book is for them.
— **David Kirtley and Daniel Bailey** for SkepticalScience.com

Bennett's careful and question-by-question presentation will lead any fair-minded person to see the warming issue more clearly and increase understanding of the need for concern.
— **Hon. George P. Shultz**, Secretary of State under President Ronald Reagan

A Global Warming Primer delivers on its promise. In engaging, accessible, and accurate prose, Jeffrey Bennett clearly explains the science of climate change, ending with a thoughtful exploration of ways to solve the problems it poses for our future.
— **Ann Reid**, Executive Director, National Center for Science Education

[*A Global Warming Primer*] covers all the basic issues of our changing climate in an easy-to-read question-and-answer format. The questions match those I have been asked many times [and the answers are presented] with clear and accurate scientific explanations that are aimed at the educated layperson, treating the reader with respect.
— **Keith Seitter**, Executive Director, American Meteorological Society

If your students are as inquisitive as mine when it comes to climate change, this book would be a great addition to your classroom. Whether used as your main text, supplemental text, or reference, *A Global Warming Primer* will provide the facts and evidence needed to teach your students effectively.
— **NSTA Recommends**

There are many books on the market right now about global warming and the implications for our planet, but rarely are they this engaging and easy to understand. This book deserves a place in middle school and high school libraries and would be a valuable resource to anyone looking to extend their knowledge about the science of our changing planet.
— ***Green Teacher Magazine***

Jeffrey Bennett has done what many others have been unable to do: He has made climate science understandable. That is a considerable achievement given the complexity of the topic and the need for all of us to grasp the basics of what is arguably the most important topic of our time.
— **Gov. Bill Ritter, Jr.** (Colorado), Author of *Powering Forward*

This delightfully perceptive book is a must-read for everyone concerned about our future. It covers climate's complex topics in a clear, illuminating manner. The insightful approach makes the subject accessible to newcomers and brings a fresh perspective that should interest even climate experts.
— **William Gail**, Past President, American Meteorological Society

This great book addresses common skeptical climate arguments in a way that sorts out the science from the belief on both sides of the debate. I also love its optimistic focus on climate solutions.
— **Piers Forster**, IPCC Lead Author and Director of the Priestley International Centre for Climate, University of Leeds, UK

A friendly yet authoritative look at how we know what we know about the climate, and why we need to do something about it.
— **Carl Zimmer**, Author of *Evolution: Making Sense of Life*

A Global Warming Primer is an exceptionally valuable resource for educators at all levels. The scientific understanding of modern global warming as well as discussion of real-world solutions are made readily accessible via the book's conveniently indexed Question and Answer format coupled with an optional, deeper tier of explanation and evidence.
— **Dr. Cherilynn Morrow**, Founder of ArtSciencEducation and Recipient of the American Geophysical Union SPARC Award for Education and Public Outreach

Concise, crystal clear, packed with the best available information — this is the book to grab if you want to be well informed about climate change. Carefully stepping around controversial politics, this "primer" will give you in an afternoon everything you need to know about the science and economics that will govern the future of our civilization.
— **Spencer Weart**, Author of *The Discovery of Global Warming*

Those of us who are curious and concerned about climate change will not find a more lucid explanation of climate science. With clear explanations and surprisingly simple examples, Jeffrey Bennett takes the intimidation factor out of what the great majority of climate scientists have been trying to explain to us for decades.
— **William Becker**, Executive Director, Presidential Climate Action Project

A remarkably clear explanation of the causes and effects of global warming and what we can do to address it.
— **David Bookbinder and David Bailey**, Element VI Consulting

As an entrepreneur working to provide sustainable real food to communities around the world, the dangers of climate change are never far from my thoughts. For anyone who doubts the reality of the threat, this is the book to read to help you understand it. Best of all, you'll come away realizing that the problem is eminently solvable, and that the solution will help create a stronger economy and better world for our children and grandchildren.
— **Kimbal Musk**, Entrepreneur, Venture Capitalist, and Co-Founder of The Kitchen

From science to solutions, this clearly written and up-to-date survey of human-caused climate change illuminates one of the great existential issues of our time.
— **Prof. Richard C. J. Somerville**, University of California, San Diego and Author of *The Forgiving Air: Understanding Environmental Change*

A creative and remarkably accessible summary of climate science and policy. Quick and easy as 1-2-3!
— **Yoram Bauman**, Ph.D., the "Stand-up Economist" and Coauthor of *The Cartoon Introduction to Climate Change*

I recommend this book to business leaders not only to better understand the threats posed to our economy by global warming, but also to appreciate the enormous business opportunities inherent in the transition to clean energy.
— **Nicole Lederer**, Founder, Environmental Entrepreneurs (E2.org)

Author Jeffrey Bennett has an extraordinary ability to explain climate change and its impact on our planet and its inhabitants in a clear and simple manner. This is the book to read to get the true story.
— **Ben Bressler**, Founder, Natural Habitat Adventures

A book that everyone should read, whether a skeptic or a believer. Dr. Bennett clearly lays out both the arguments and the explanations, using a perfect mix of answering tough questions in a simple and straightforward way that everyone can understand, along with further background for those wanting a more detailed scientific explanation. I applaud him for taking on such a difficult but critically important topic in a way that will inspire everyone to think more hopefully about the question: What can *I* do about it?
— **Dr. Susan Lederer**, NASA Space Scientist

A very readable and understandable presentation of the basic science of global warming. This book could go a long way in creating an informed electorate on one of the most important issues of our day.
— **Stephen Turcotte**, Professor of Physics, BYU-Idaho

A must read for everyone on our planet. *A Global Warming Primer* presents the facts while leaving the politics out in a way non-scientists can clearly understand. We hope everyone will do what he suggests and write "A Letter to Our Grandchildren" about what we as individuals will do based on the evidence.
— **Mark Levy and Helen Zentner**, Educational Consultants

A Global Warming Primer is what you get when a first-rate writer and educator brings his cosmic perspective to the most pressing issue facing humanity. Bennett's Q&A approach, while easily digestible, is rooted in complex science that few can relate so simply, clearly, and readably. Highly recommended.
— **Todd Neff**, Science Writer and Author of *From Jars to Stars*

This book is just what I've been looking for as a teaching aid and primer on this subject. It elegantly summarizes a huge amount of complex information and speaks with an authoritative voice clearly based on years of experience of teaching and writing on the subject.
— **James McKay**, Editor of *Dreams of a Low Carbon Future* and Manager of the Centre for Doctoral Training in Low Carbon Technologies, University of Leeds, UK

This book offers us just what we need right now: clarity. Step by step, question by question, the author states the facts, explains the underlying concepts, and offers us the best comfort we can have: the power to honestly face the facts of our changing planet. I wish everyone in the world would pause and read this book.
— **Dr. Michelle Thaller**, Astronomer and TEDx speaker

A Global Warming Primer takes a complicated topic and breaks it down in a simple way that anyone can understand while also bridging the partisan divide that sometimes gets in the way of the science. There are multiple references throughout for those who want to delve deeper into the topics, but the general format is focused and concise, making it a quick and easy read. Everyone should read this important text — our future and our children's future depend on it.
— **Gabe L. Finke**, CEO, Ascentris

Who better to help us understand global warming than the astrophysicist and educator Jeffrey Bennett! Love how this book walks us through the scientific facts and addresses skeptic claims to provide us with intelligent talking points for discussions with families, friends, and co-workers on this important world issue.
— **Patricia Tribe**, CEO, Story Time From Space, and Former Director of Education for Space Center Houston

Bennett cuts right through the noisy arguments about climate change and shows that global warming is an inevitable consequence of simple physics and the fuels we burn. In his marvelously accessible style, Bennett tells it like it is. If you don't know what to think about climate change — or even if you do — this is the one book to read.
— **Seth Shostak**, SETI Institute and Host of *Big Picture Science*

I have read dozens and dozens of climate books and can say without equivocation that *A Global Warming Primer* should be on your short list. From science to solutions, Jeffrey Bennett provides comprehensive information in an easily understandable style.
— **Scott Mandia**, Professor of Physical Sciences, Suffolk County Community College

We collectively owe author Jeffrey Bennett a huge "Thank You" for this effort to enlighten anyone with questions about global warming. He has a remarkable ability to communicate complicated atmospheric and oceanic climate factors in a manner that will permit nonscientist readers to comprehend and appreciate the critical importance of this topic.
— **Ron Alberty**, Former Chief of Meteorological Research, National Severe Storms Laboratory

A Global Warming Primer fills a unique niche: providing a very clear and accessible description of what we know about climate change, where there are uncertainties, and the range of possible solutions. We live in a world where people's understanding of climate change is often correlated with their political beliefs — hopefully this primer will help create a fact-based understanding of the underlying science and the choices before us.
— **Dr. Will Toor**, Executive Director, Colorado Energy Office

With clear and detailed explanations, scientist and educator Jeffrey Bennett carefully dismantles the misconceptions that have clouded the debate on climate change, then presents the solutions we must pursue to solve this critical challenge. This book is a must-read for believers and skeptics alike.
— **Andrew Chaikin**, Author of *A Man on the Moon*

For reasons that are unreasonable to reasonable people, climate change science — and science more broadly — has been argued in recent times as a partisan subject. But this science is not blue or red; it's simply science that is essential for humankind to understand. This book admirably helps us do just that.
— **Sven Lindblad**, President and CEO, Lindblad Expeditions, Inc.

I'm not a scientist, and I didn't need to be to understand this book. By presenting the evidence-based facts simply and clearly, this book will enable anyone to understand why global warming is an issue that we can't afford to ignore.
— **RJ Harrington, Jr.**, President and CEO, Sustainable Action Consulting

In clean, clear, elegant, and engaging style, Dr. Bennett lays out what scientists know about climate change and explains it in a way that will enable both young people and adults to understand what we know and how we know it. This book is a major contribution to climate literacy, taking just the right approach to engage the reader and help us all become smarter inhabitants of home planet Earth.
— **Dan Barstow**, CASIS Education Manager and Founder, Climate Literacy Network

By sticking to the facts in an easily accessible Q&A format, Bennett deftly explores a complicated and most critical problem of our time.
— **Susan Nedell**, Rocky Mountains Advocate, Environmental Entrepreneurs (E2)

By illustrating the challenges as well as the solutions, Jeffrey Bennett's reliance on a "big picture" approach to climate change empowers his fellow citizens to embrace a fact over fear approach to resolving our climate crisis.
— **Christina Erickson**, Attorney and Environmental Advocate

A Global Warming Primer

SECOND EDITION

Also by Jeffrey Bennett

For Children
Max Goes to the Moon
Max Goes to Mars
Max Goes to Jupiter
Max Goes to the Space Station
The Wizard Who Saved the World
I, Humanity
Totality! An Eclipse Guide in Rhyme and Science

For Grownups
Beyond UFOs: The Search for Extraterrestrial Life and Its Astonishing Implications for Our Future
Math for Life: Crucial Ideas You Didn't Learn in School
What Is Relativity? An Intuitive Introduction to Einstein's Ideas, and Why They Matter
On Teaching Science: Principles and Strategies That Every Educator Should Know

Middle/High School Textbooks
Earth and Space Science — Free, online at grade8science.com

High School/College Textbooks
The Cosmic Perspective
The Essential Cosmic Perspective
The Cosmic Perspective Fundamentals
Life in the Universe
Using and Understanding Mathematics: A Quantitative Reasoning Approach
Statistical Reasoning for Everyday Life

A Global Warming Primer

SECOND EDITION

Pathway to a Post–Global Warming Future

Jeffrey Bennett

BIG KID SCIENCE

Boulder, CO

Education, Perspective, and Inspiration for People of All Ages

About the Cover

I could put my thumb up to a window and completely hide the Earth.
— Apollo 8 Astronaut Jim Lovell

The cover shows the iconic "Earthrise" photo taken during the Apollo 8 mission — the first mission to carry humans as far as the Moon — by astronaut William Anders on Christmas Eve, 1968. The image makes it immediately clear that we live on a small and precious planet, and it played an important role in building awareness of the need to preserve our environment. In a post–global warming future, we can imagine having restored the climate to what it was at the time this photo was taken. Moreover, the abundant energy and wealth of that future world would make it possible for us to share it with peace and prosperity for all — and with many more of us being able to see the Earth as it looks from the Moon and beyond.

Published by
Big Kid Science
Boulder, CO
www.BigKidScience.com

Education, Perspective, and Inspiration for People of All Ages

Book website: **www.GlobalWarmingPrimer.com**

Distributed by IPG
Order online at www.ipgbook.com
or toll-free at 800-888-4741

Editing: Lifland et al., Bookmakers
Composition and design: Side By Side Studios

ISBN: 978-1-937548-88-9

To Grant and Brooke,
in hopes that you and your generation
will live your lives in a world far better than
that of any past generation.

Brief Table of Contents

Detailed Table of Contents

Introduction — Envisioning a "Post–Global Warming" Future

Preservation of our environment is not a liberal or conservative challenge; it's common sense.

— President Ronald Reagan, Jan. 25, 1984 (State of the Union address)

Imagine for a moment that the world comes together to stop the emissions of carbon dioxide (and other greenhouse gases) that have been causing our planet to warm up, and that we then develop and implement technology to actively remove carbon dioxide from the atmosphere. With these steps, we could not only stop the current warming trend in its tracks, but also begin to restore the climate to a better and more natural state. In doing so, we would achieve what you might call a "post–global warming" future, meaning a future in which today's children will someday be able to talk about global warming as a once-serious problem that we found a way to solve. Moreover, because achieving this future would mean we had also found ways to produce clean and abundant energy, the post–global warming future could be one in which we use this energy to raise living standards around the world, to protect and preserve the global environment, and to build an era of global peace and prosperity with far greater equality for all.

This vision of a post–global warming future world might sound like fantasy, but it is no less realistic than the far more dismal climate futures that we commonly hear about in the media. We already have all the technology we need to replace our current energy sources with alternatives that not only are cleaner but also offer overall economic benefits, and we are rapidly developing new technologies that offer the promise of energy abundance far beyond what we have today. Achieving a post–global warming future simply requires that we make wise and informed choices, which brings me to the three main goals of this book:

1. The first step in solving any problem is to understand it, so my first goal is to help you understand the basic science of global warming and the evidence that underlies this science.

1

2. I also want to help you understand the consequences of global warming — including both what we are already seeing and what might occur in the future — so that you will understand the very real threat that it poses if left unaddressed.

3. I want to convince you that despite the "doom and gloom" often associated with this topic, it really is possible for us to create a post–global warming future. If we act quickly enough, this amazing future could be available to the generation of youth that is already alive today. Moreover, because this future is one in which our economy will be stronger and energy will be cheaper, cleaner, and more abundant than it is today, it should be possible for us to find a pathway to this future that people of all political persuasions can agree on.

To accomplish these goals, I've structured this book with five main chapters. In chapter 1, I will focus on the basic science of global warming, emphasizing what I call the "1-2-3" logic that underlies the subject. In chapter 2, I'll address common "skeptic" claims that have sometimes been leveled against the science of global warming, a task that will also give us the opportunity to explore a few important details of climate science. Chapter 3 will focus on the major consequences of global warming, which I will divide into five major categories to make it easy for you to understand and remember them. I'll devote chapter 4 to the solutions that, at least in principle, could help us build a pathway to a post–global warming future. The short chapter 5 is designed to help you think on a personal level about what it will take to put us on that pathway.

As you will see, much of this book uses a question-and-answer format, with questions drawn from the many that I've been asked over the years. I hope that this format will make the book feel at least a little more like a personal discussion. You'll also notice that the book uses two distinct font sizes. The normal font is for general text (like this introduction) and the big picture ideas that should be of interest to all readers, while this smaller font is for more detailed discussion that you can treat as optional, depending on the depth to which you'd like to go.

Finally, a quick note about the changes I've made to this book from its first edition, which was published back in 2015. First, I have updated all the scientific data and discussions. Second, I've revised the Q&A set to reflect new questions that I've heard while speaking on this topic. Third and most important (and reflected in the new subtitle), I've subtly changed the general perspective with which I approach the topic based on a fairly dramatic change that has occurred in public perception. At the time of the first edition, I felt that the biggest challenge in talking about global warming lay in convincing people that the threat was real. However, polls now show far more acceptance of the reality of climate change (probably because of the many extreme weather events that have been occurring), and the biggest challenge has become one of combatting the despair that many people — especially young people — now feel about their climate future.

It is my fervent hope that this book will help overcome that despair by demonstrating that as serious as the threat really is, we have the power to create a future as bright as any that you might imagine. All it takes is for all of us to work together, and to appreciate the simple truth embodied in the opening quote from Ronald Reagan: When it comes to preserving the environment in which we all live, we just need a little common sense.

Jeff Bennett
Summer 2023

The Basic Science —
Easy as 1-2-3

What we are now doing to the world . . . by adding greenhouse gases to the air at an unprecedented rate . . . is new in the experience of the Earth. It is mankind and his activities which are changing the environment of our planet in damaging and dangerous ways.

— British Prime Minister Margaret Thatcher, Nov. 8, 1989 (speech to the United Nations)

We all know that human activities are changing the atmosphere in unexpected and in unprecedented ways.

— President George H. W. Bush, Feb. 5, 1990 (remarks to the Intergovernmental Panel on Climate Change)

The pathway to a post–global warming future begins with understanding the nature of the problem that we face today. You might think that this would be difficult given the extreme complexity of Earth's full climate system, but the underlying science of global warming turns out to be remarkably simple. I'll use this first chapter to describe this basic science and the evidence that backs it up.

Before we begin, let me draw your attention to the two quotes above. Notice that both are from conservative leaders, both are from more than 30 years ago, and both reflect clear certainty about the reality and threat of global warming. As you may have guessed, I chose these quotes to emphasize that the political polarization that often accompanies this topic represents a false and relatively recent narrative. The science itself is as clear as can be. So if you have any doubts of your own about the science or have friends or family who express doubts, I hope that this chapter will help to dispel them.

A Tale of Two Planets

Figure 1.1 shows the planets Earth and Venus to scale, along with their global average surface temperatures. You can see that both planets are about the same size, and from spacecraft studies and other data we know that they also both have about the same overall mass, density, and composition. Indeed, in most respects Venus and Earth are about as close to being identical twin planets as it is probably possible for planets to be, with one striking exception: their average surface temperatures.[1] Earth has an average surface temperature ideally suited to life and our civilization, while Venus has a surface temperature hot enough to melt lead. From a scientific perspective, this raises a fairly obvious question: Why would two otherwise similar planets have such drastically different surface temperatures?

It might be tempting to chalk the temperature difference up solely to the fact that Venus is closer to the Sun, but a little further thought tells us that this cannot be the full story. Figure 1.2 shows part of the Voyage Scale Model Solar System, which depicts the sizes and distances of the Sun and planets on a scale of 1 to 10 billion. Notice that while it's true that Venus is closer to the Sun, the difference isn't really all that great, and calculations confirm that it's not nearly enough to account for such a large temperature difference by itself. Moreover, Venus's bright clouds reflect so much sunlight that its surface actually absorbs *less* sunlight than Earth's, which by itself would lead us to expect Venus to be *colder* than Earth. So why is Venus so hot?

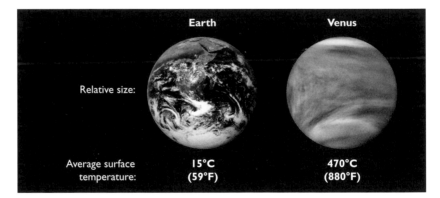

Figure 1.1. Earth and Venus are shown to scale. Why do two planets that are otherwise so similar have such different surface temperatures?

---

1   In science, temperatures are almost always stated on the Celsius (°C) scale, which is also the temperature system used in most of the world. In this book, I will generally also provide Fahrenheit (°F) equivalents, since those are more familiar to readers in the United States.

Figure 1.2. This photo shows the inner portion of the Voyage Scale Model Solar System, located outside the National Air and Space Museum (Washington, D.C.). The locations of the Sun, Venus, and Earth are indicated. The model Sun is the visible gold sphere. Venus and Earth are each about the size of the ball point in a pen (about one millimeter in diameter) on this scale and can be seen within the glass disks that face outward from their pedestals.

Note: Voyage models are now also found in many other communities across the United States; visit voyagesolarsystem.org to learn more, including how you can get a Voyage model for your own campus or community.

Scientists have investigated virtually every possible explanation, and there is only one that works: The huge difference between the temperatures of Venus and Earth arises primarily from differences in their amounts of atmospheric carbon dioxide, a gas that can trap heat and make a planet warmer than it would be otherwise. In fact, as we'll discuss in more detail shortly, both planets would actually be frozen over if they had no carbon dioxide in their atmospheres at all. Earth has just enough carbon dioxide (plus water vapor; see page 16) to make our planet livable, so in that sense carbon dioxide is a very good thing for life. But Venus has almost 200,000 times as much carbon dioxide in its atmosphere as Earth, and all this carbon dioxide traps so much heat that the entire surface is baked hotter than a pizza oven — providing clear proof that it is possible to have too much of a good thing (figure 1.3).

This story of Venus and Earth contains almost everything you need to understand the basic science of global warming. It shows that gases like carbon dioxide, which we call *greenhouse gases*, really do make

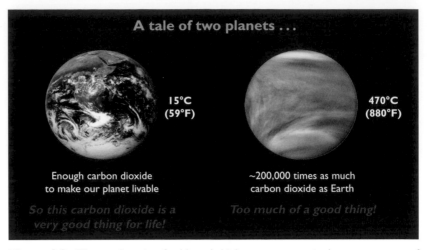

Figure 1.3. The explanation for Venus's high temperature is the vast amount of carbon dioxide in its atmosphere, which traps an enormous amount of heat through what we call the *greenhouse effect*.

planets warmer than they would be otherwise, and that the more of these gases a planet has, the hotter it will be.

Global Warming 1-2-3

The lesson from our tale of two planets leads directly to the subtitle of this chapter, in which I stated that understanding global warming is as easy as 1-2-3. By this I mean that despite the fact that Earth's complete climate system is very complex, the basic science of global warming can be summarized in three simple statements, which embody two indisputable scientific facts and the inevitable conclusion that follows from them:

1. ***Fact:*** Carbon dioxide is a greenhouse gas, by which we mean a gas that traps heat and makes a planet (such as Earth or Venus) warmer than it would be otherwise.
2. ***Fact:*** Human activity, especially the use of fossil fuels[2] — by which we mean coal, oil, and natural gas, all of which release carbon dioxide when burned — is adding significantly more of this heat-trapping gas to Earth's atmosphere.

---

2   Fossil fuels (coal, oil, gas) get their name from the fact that they come from the remains (fossils) of living organisms that died and decomposed long ago. They are rich in carbon because all life on Earth is based on carbon. When they burn, their carbon combines with oxygen to produce carbon dioxide. Overall, about 90% of our carbon dioxide emissions are from the use of fossil fuels.

3. *Inevitable Conclusion:* Given that more carbon dioxide means warmer temperatures and that we are adding carbon dioxide (and other greenhouse gases) to Earth's atmosphere, it is inevitable that global warming should occur as a result. The more of this gas we add, the greater the warming will be.

In the rest of this chapter, we'll discuss the evidence establishing that the first two facts are true, along with the evidence showing that the expected warming is indeed under way.

Evidence for Fact I (Carbon Dioxide Makes Planets Warmer)

Fact 1 is that carbon dioxide is a greenhouse gas that makes a planet warmer than it would be otherwise. Now, in Q&A format, we're ready to examine the evidence that makes this a fact rather than a matter of opinion.

How do we know that Fact I is really a fact?

There is no doubt that higher concentrations of carbon dioxide and other greenhouse gases make planets warmer, because this fact is based on the simple, well-understood, and well-tested physics of what we call the *greenhouse effect*. Figure 1.4 shows how the greenhouse effect works. Notice the following key ideas:

* The energy that warms Earth comes from sunlight, and in particular from visible light (the kind of light that our eyes can see). Some sunlight is reflected back to space, and the rest is absorbed by the surface (land and oceans).
* Earth returns the energy it absorbs back upward, but not in the form of visible light (if it did, Earth would glow in the dark). Instead, the returned energy takes the form of *infrared light*, which our eyes cannot see.
* Greenhouse gases — which include water vapor (H_2O), carbon dioxide (CO_2), and methane (CH_4, the predominant ingredient of natural gas) — are made up of molecules[3] that are particularly good at absorbing infrared light. Each time a greenhouse gas mol-

---

3 Recall that all ordinary matter is made up of atoms, but sometimes atoms are bound together in *molecules*. We state molecular composition with simple formulas, like H_2O, which means a molecule made up of two hydrogen (H) atoms and one oxygen (O) atom. Similarly, carbon dioxide is CO_2 because it consists of molecules with one carbon (C) atom and two oxygen (O) atoms, while methane is CH_4 because it has one carbon (C) atom and four hydrogen (H) atoms.

The Greenhouse Effect

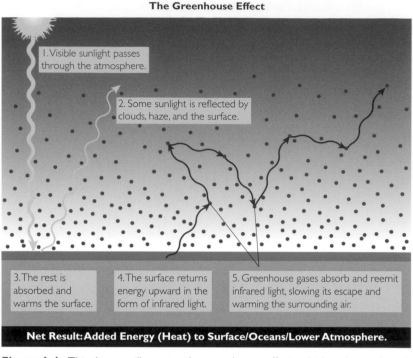

1. Visible sunlight passes through the atmosphere.

2. Some sunlight is reflected by clouds, haze, and the surface.

3. The rest is absorbed and warms the surface.

4. The surface returns energy upward in the form of infrared light.

5. Greenhouse gases absorb and reemit infrared light, slowing its escape and warming the surrounding air.

Net Result: Added Energy (Heat) to Surface/Oceans/Lower Atmosphere.

Figure 1.4. This diagram illustrates the greenhouse effect, which makes a planet's surface, oceans, and lower atmosphere warmer than they would be otherwise. The yellow arrows represent visible light, the red arrows represent infrared light, and the blue dots represent greenhouse gas molecules. The more greenhouse gas (blue dots in this figure), the greater the warming.

ecule absorbs a photon (the technical name for a "piece" of light) of infrared light, the absorbed energy goes into molecular motion (heat) and to reemitting one or more other infrared photons, which may head off in any random direction. These photons can then be absorbed by other greenhouse gas molecules, which do the same thing.

The net result is that greenhouse gases tend to slow the escape of infrared light from the lower atmosphere, while their molecular motions heat the surrounding air. In this way, the greenhouse effect makes the surface, oceans, and lower atmosphere warmer than they would be from sunlight alone. The more greenhouse gases present, the greater the warming. A blanket offers a good analogy. You stay warmer under a blanket not because the blanket itself provides any heat, but because it slows the escape of your body heat into the cold outside air. The more blankets (or greenhouse gases in our analogy), the warmer you will be.

Q **How do we know that Earth returns energy to space in the form of infrared light?**

It's basic physics, verified by observations. All objects — including the Sun, the planets, and even you — always emit energy in the form of light,[4] but the specific type of light depends on the temperature. Hot objects, like the Sun, can emit visible light. Cooler objects, like planets and you, emit only infrared light. While we cannot see infrared light with our eyes, we can detect it with infrared cameras and other instruments, and orbiting satellites can directly measure the amount of infrared light being emitted by Earth. In fact, if you look ahead to page 70, you'll see that these types of satellite measurements also tell us how much extra energy is being trapped as the greenhouse effect strengthens.

Q **Why haven't you mentioned nitrogen and oxygen, which make up most of our atmosphere?**

The atmosphere is indeed made mostly of nitrogen and oxygen. Together, these two gases make up about 99% of Earth's "dry" atmosphere (78% for nitrogen and 21% for oxygen), meaning the atmospheric composition in the absence of water vapor. However, molecules of nitrogen and oxygen do not absorb infrared light, and therefore do not contribute to the heating of the surface. In other words, without the relatively small amounts of infrared-absorbing greenhouse gases (especially water vapor, carbon dioxide, and methane) that are present in our atmosphere, all the infrared light emitted from Earth's surface would escape directly into space, and our planet would be frozen over.

In case you are wondering why some molecules can absorb infrared light and others cannot, it is a result of their structures. In our atmosphere, nitrogen and oxygen are molecules in which two atoms are bound together; that is, nitrogen is in the form N_2 and oxygen in the form O_2. In order to absorb photons of infrared light, molecules must be able to vibrate and rotate in ways that distort their shapes. This turns out to be fairly difficult for molecules with only two atoms, particularly when both atoms are the same, as in N_2 and O_2. In contrast, these types of vibration and rotation are relatively easy for many molecules with more than two atoms, which is why water vapor (H_2O), carbon dioxide (CO_2), and methane (CH_4) all absorb infrared light effectively, making them greenhouse gases.

Q **Are there any other greenhouse gases I should know about?**

Although water vapor, carbon dioxide, and methane are the three most important greenhouse gases in Earth's atmosphere, a few other atmospheric gases can also act as greenhouse gases and therefore contribute to warming. Those that we'll discuss a bit more in this book are nitrous oxide (N_2O) and

---

4 More technically, objects generally emit what we call *thermal radiation* (also called *blackbody radiation*), which has a characteristic spectrum in which the emitted light has an intensity and range of wavelengths that depend only on the object's temperature. Hotter objects have spectra that peak at shorter wavelengths, while also emitting more energy per unit area at all wavelengths than cooler objects.

industrial chemicals known as halocarbons, which include chlorofluoro-carbons (CFCs).

Q **I've heard that "greenhouse effect" is a misnomer. Is that true?**

It depends on how picky you want to be. The term comes from botanical greenhouses, but those greenhouses actually trap heat through a different mechanism than planetary atmospheres: Rather than absorbing infrared radiation, greenhouses stay warm largely by preventing warm air from rising. Nevertheless, atmospheric greenhouse gases and botanical greenhouses have the same net effect of making things warmer than they would be otherwise, so I'm personally fine with the term "greenhouse effect."

Q ## How do we know that greenhouse gases really trap heat?

Two major lines of evidence show conclusively that greenhouse gases trap heat. First, scientists can measure the heat-trapping effects of these gases in the laboratory. This type of work was first reported way back in 1856 by the American scientist Eunice Newton Foote, who discovered that carbon dioxide could absorb heat from sunlight and suggested that variation in the atmospheric concentration of carbon dioxide might therefore affect the climate.[5] Soon thereafter, British scientist John Tyndall began measuring the precise greenhouse effects of a variety of gases. Although the actual setup was somewhat more complex (figure 1.5), the basic idea is simply to put a gas (such as carbon dioxide) in a tube, shine light of different wavelengths (such as visible and infrared) at it, and measure how much of that light passes through and how much is absorbed. By the end of the 19th century, such measurements were sophisticated enough that a famous chemist, Svante Arrhenius, used them to calculate the expected effect of a doubling of Earth's carbon dioxide level. His result, a temperature increase of about 5°C (9°F), was remarkably close to what the most sophisticated models predict today.

The second line of evidence extends these laboratory measurements to the real world — or plural *worlds* in this case — confirming that the greenhouse effect raises actual planetary temperatures as expected. If there were no greenhouse effect, a planet's average temperature would depend only on its distance from the Sun and the relative proportions of sunlight that it absorbs and reflects. I won't bother you with the mathematical details, but they lead to the simple formula that you can see being applied to Earth in figure 1.6. The formula shows that Earth's

---

5 Foote's pioneering work was presented at an 1856 scientific conference by a male colleague, Joseph Henry. Although Henry credited Foote, her work did not become widely known and was only recently "rediscovered" by historians of science. As a result, credit for the first work on the heat-trapping effects of gases was long given to John Tyndall — a mistake I made myself in the first edition of this book. Note, however, that Tyndall was the first to measure what we now call the greenhouse effect, meaning *infrared* absorption, as Foote focused on the effects of sunlight.

The Basic Science

John Tyndall's Setup to Measure Light (1859)

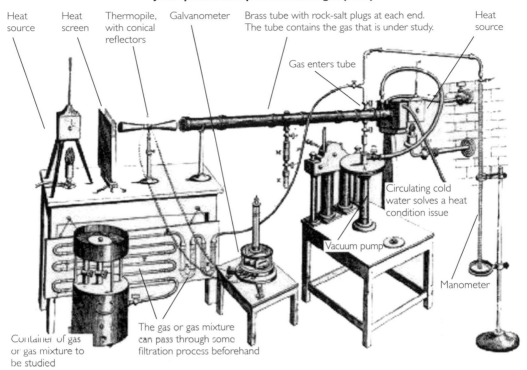

Heat source

Heat screen

Thermopile, with conical reflectors

Galvanometer

Brass tube with rock-salt plugs at each end. The tube contains the gas that is under study.

Heat source

Gas enters tube

Circulating cold water solves a heat condition issue

Vacuum pump

Manometer

Container of gas or gas mixture to be studied

The gas or gas mixture can pass through some filtration process beforehand

Figure 1.5. This diagram shows the experimental setup used by John Tyndall in 1859 to measure how gases like carbon dioxide create what we now call the greenhouse effect. The measurements have been repeated and refined ever since.

Source: The original illustration is from Tyndall's 1872 book *Contributions to Molecular Physics in the Domain of Radiant Heat*; this annotated version is from Wikipedia.

global average temperature would be well below freezing (–16°C, or +3°F) without greenhouse gases. In other words, we need the green house effect to explain Earth's actual average temperature of about 15°C (59°F). The same is true for all other planets (as well as for Saturn's moon Titan, the only moon with a substantial atmosphere): We get correct answers for their temperatures only when we use mathematical formulas that include the greenhouse effect.

This brings us back to our tale of two planets. We find that without the naturally occurring greenhouse effect, Earth would be too cold for liquid oceans and life as we know it. That is why, as we saw in figure 1.3, the natural greenhouse effect is a very good thing for life on Earth. But Venus, with almost 200,000 times as much carbon dioxide in its atmosphere as Earth, clearly has much too much of this good thing.

Q Why does Venus have so much carbon dioxide in its atmosphere?

Earth actually has about the same total amount of carbon dioxide as Venus, but while Venus's carbon dioxide is virtually all in its atmosphere, nearly all of Earth's is "locked up" in what we call *carbonate rocks*, the most familiar of

Figure 1.6. This painting depicts the calculation of Earth's expected average temperature if there were no greenhouse effect. The fact that this temperature (−16°C) is so much lower than the actual average temperature (+15°C) shows that the natural greenhouse effect is what makes Earth warm enough for life.

Painting by Roberta Collier-Morales from *The Wizard Who Saved the World*.

which is limestone. The reason for this difference is that Earth has oceans and Venus does not.

On both planets, the original source of carbon dioxide was gas released by volcanoes. On Earth, carbon dioxide dissolves in the oceans (which contain about 60 times as much carbon dioxide as the atmosphere), where it then combines with dissolved minerals to form carbonate rocks (which contain almost 200,000 times as much carbon dioxide as the atmosphere). Venus lacks oceans and therefore cannot dissolve carbon dioxide gas, so it all remains in the atmosphere.

A deeper question is why Earth has oceans and Venus does not, and here Venus's closer distance to the Sun *does* make a difference. You can understand why by thinking about what would happen if Earth were magically moved to Venus's orbit (figure 1.7). The greater intensity of sunlight would immediately raise Earth's average temperature from its current 15°C to about 45°C (113°F). The higher temperature would increase the evaporation of water from the oceans, putting much more water vapor into the atmosphere — and because water vapor is a greenhouse gas, the added water

If Earth moved to Venus's orbit

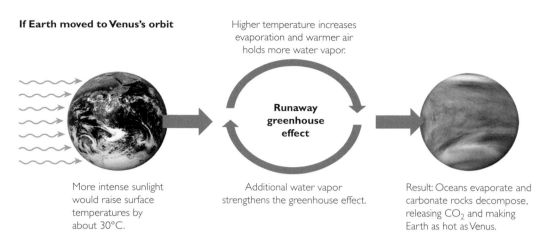

Higher temperature increases
evaporation and warmer air
holds more water vapor.

Runaway greenhouse effect

More intense sunlight
would raise surface
temperatures by
about 30°C.

Additional water vapor
strengthens the greenhouse effect.

Result: Oceans evaporate and
carbonate rocks decompose,
releasing CO_2 and making
Earth as hot as Venus.

Figure 1.7. This diagram shows what would happen if Earth magically moved to Venus's orbit.

vapor would strengthen the greenhouse effect and drive temperatures even higher. The higher temperatures, in turn, would lead to even more ocean evaporation and more water vapor in the atmosphere, strengthening the greenhouse effect even further. In other words, we'd have a *reinforcing feedback*[6] process in which each little bit of additional water vapor in the atmosphere would lead to a higher temperature and even more water vapor. The process would rapidly spiral out of control, resulting in what scientists call a *runaway greenhouse effect*. It would not stop until the "moved Earth" became as hot as (or even hotter than) Venus is today.

In fact, something like this probably occurred on Venus long ago. Based on scientific understanding of how the Sun generates energy through nuclear fusion, the Sun should very gradually brighten with time. The rate is so slow that we cannot measure it, but calculations indicate that the Sun was about 30% dimmer when the planets were born (about 4½ billion years ago) than it is today. This means that the young Venus probably had sunlight of not much greater intensity than Earth does today, and some scientists suspect that Venus may have had oceans at that time. As the Sun gradually brightened, Venus grew hotter until a runaway greenhouse effect set in.

Q Wait — I heard a claim that it is Venus's high pressure (rather than the greenhouse effect) that explains Venus's high temperature; could that be true?

This claim has indeed circulated in some corners of the web, but it is easy to understand why it is incorrect. The surface pressure on any planet is determined by the total weight of atmospheric gas above the surface, so Venus's high surface pressure is a direct result of the huge amount of gas — almost all carbon dioxide — that makes up its atmosphere. The pressure created by the enormous weight of this gas would be the same whether the gas were

---

6 In science, a reinforcing feedback is more commonly called a *positive feedback*; it is a feedback that adds to and thereby amplifies an existing effect, in contrast to a self-regulating, or *negative,* feedback, which reduces an existing effect.

hot or cold; the only difference would be that the atmosphere would be more "puffed up" if it were hotter.[7] So pressure alone does not determine temperature, and as discussed above, we can correctly predict Venus's temperature only by taking into account the greenhouse effect caused by the carbon dioxide.

Q **Does Mars also have a greenhouse effect?**

Yes, but it is very weak. The atmosphere of Mars is made mostly of carbon dioxide (about 95%), but the atmosphere is so thin (the surface pressure is less than 1% of that on Earth) that the total amount of carbon dioxide is actually quite small. As a result, Mars is warmed only a little by its greenhouse effect, and its greater distance from the Sun makes it quite cold, with an average surface temperature of –50°C (–58°F). Scientifically, the surprise is that Mars shows clear evidence of having had liquid water on its surface in the past. This means that long ago Mars must have been much warmer than it is today, which in turn means that it must once have had a much stronger greenhouse effect. Scientists have a pretty good idea of why Mars once had a strong greenhouse effect and why the effect ultimately weakened so much, but the full discussion isn't directly relevant to our topic in this book.[8]

Q **Why are you focusing on carbon dioxide, when there's more water vapor (also a greenhouse gas) in Earth's atmosphere?**

It's true that there's more water vapor than carbon dioxide in Earth's atmosphere. In fact, there's about 10 times as much water vapor as carbon dioxide, and water vapor contributes more than carbon dioxide to Earth's overall greenhouse warming. However, *carbon dioxide is the more critical gas in setting Earth's temperature.*

The reason is that once we increase the carbon dioxide concentration of the atmosphere, the concentration remains enhanced for many decades, centuries, and even millennia. In contrast, water vapor cycles easily into the atmosphere through evaporation and out of the atmosphere through rain and snow. As a result, the amount of water vapor in the atmosphere at any given time is determined by the temperatures of the ocean and atmosphere. That is, the amount of water vapor in the atmosphere changes in response to temperature changes, but it does not initially cause those changes. Instead — and very importantly — water vapor *amplifies* climate changes initiated by other factors, because it acts as a reinforcing feedback. For example, if more carbon

---

7   This idea is expressed more technically through the *ideal gas law*, which can be written as $P = nkT$, where P is pressure, n is the number density (number of gas molecules per unit volume), and T is temperature; k is a fixed number known as "Boltzmann's constant." The fact that density and temperature are both on the same side of the equation means that, for a given pressure, you can have either high density and low temperature or low density and high temperature.

8   You can find many sources for learning more about our current scientific understanding of Mars and how its climate changed, including discussions in my astronomy and astrobiology textbooks (*The Cosmic Perspective* and *Life in the Universe*).

dioxide raises the global temperature a little bit, the atmosphere can hold more water vapor, which then traps more heat, making the temperature rise even more. Conversely, if the carbon dioxide level drops, the global temperature decreases, so there is less water vapor in the atmosphere, which then traps less heat, so the temperature drops further. This amplification by water vapor is well understood and is necessary to explaining Earth's natural cycles of ice ages and warm periods, which we'll discuss in chapter 2.

A bicycle makes a good analogy if you think of water vapor as the wheels and carbon dioxide as the pedals (and crank): It is the spin rate of the wheels that determines how fast the bicycle goes, but their spin rate is determined by how you drive them by pedaling. In much the same way, changes in the amount of carbon dioxide in the atmosphere drive temperature changes, even though most of the total temperature change comes from the changes in water vapor that occur as a result.

Q Can you be more precise about how long added carbon dioxide remains in the atmosphere?

Yes, but we also have to phrase the question more precisely. Let's pose it this way: If we suddenly stopped adding carbon dioxide to the atmosphere, how long would it take for the carbon dioxide concentration to drop back down to something closer to its "natural" (preindustrial) value? To answer this question, scientists consider their understanding of the many ways in which carbon dioxide is removed from the atmosphere, which include uptake by plants, dissolving in the oceans, and the gradual production of seashells and carbonate rocks. The details are fairly complex and subject to some uncertainties, and the answer also depends on how much carbon dioxide we've added (in total) by the time we stop adding it, but here's a brief summary of current understanding:

For the first few decades, uptake by the land and oceans would remove carbon dioxide relatively rapidly, so about a third of all the carbon dioxide we'd added to the atmosphere since preindustrial times would be removed in 20 to 50 years. But then the rate would slow dramatically. Even after 2,000 years, between 15% and 40% of our added carbon dioxide would still remain in the atmosphere, and it would take tens of thousands of years for the carbon dioxide concentration to come all the way back down to its preindustrial value.[9] (This discussion assumes natural processes only, as opposed to the possible future use of technologies that can remove carbon dioxide from the atmosphere.)

Q **What about other greenhouse gases?**

After carbon dioxide and water vapor, methane (CH_4) is the next most abundant greenhouse gas in Earth's atmosphere. Though it is much less abundant than carbon dioxide and makes a smaller overall impact

---

9 A good summary of these ideas is presented in the IPCC Working Group 1 Report (2013), Chapter 6, FAQ 6.2, which can be downloaded from www.ipcc.ch/report/ar5/wg1/.

on the climate, its effect is still important. The same is true for other greenhouse gases, and scientists can and do take all these greenhouse gases into account when gauging the strength of the greenhouse effect. However, to keep our discussions simple, I will generally focus on carbon dioxide in this book, since it has the greatest impact on the climate. Nevertheless, the contribution of other greenhouse gases is very important to consider both scientifically and in policy decisions, so while they may be secondary in importance to carbon dioxide, they should not be ignored. (See the brief discussion of these other greenhouse gases on pages 27–28).

What's the bottom line for Fact 1 (carbon dioxide makes planets warmer)?

We have seen overwhelming evidence that Fact 1 really is a fact. The heat-trapping effects of greenhouse gases have been measured in the laboratory, and our understanding of the greenhouse effect has been further verified by the fact that actual planetary temperatures match calculated temperatures only when we take it into account. We are therefore left with no scientific doubt that carbon dioxide and other greenhouse gases in a planet's atmosphere create a greenhouse effect that makes the planet warmer than it would be otherwise.

Evidence for Fact 2 (Human Activity Is Adding Carbon Dioxide to the Atmosphere)

We now turn to the evidence that establishes Fact 2, which is that human activity, especially the use of fossil fuels, is adding heat-trapping carbon dioxide to Earth's atmosphere.

How do we know that the atmospheric concentration of carbon dioxide is really rising?

The most direct way to measure the amount of carbon dioxide in the atmosphere is to collect and study air samples. Scientists have been making such direct measurements continuously since the late 1950s. Figure 1.8 displays the measurements made using samples collected at the Mauna Loa Observatory in Hawaii. As you can see, the measurements clearly show a rapidly rising concentration of carbon dioxide in Earth's atmosphere. Measurements made at numerous other sites around the world confirm this increase over time.

Notice that the units on the vertical axis are *parts per million* (ppm), giving the number of carbon dioxide molecules among each 1 million total molecules of air. You can see that the carbon dioxide concentra-

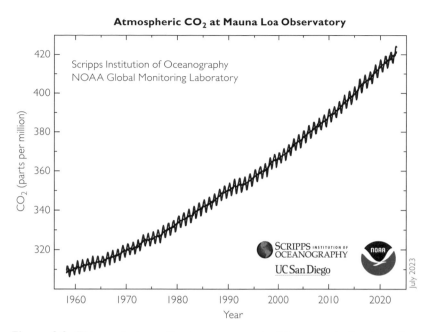

Figure 1.8. This graph shows direct measurements of the amount of carbon dioxide in the atmosphere, which have been made on a regular basis since the late 1950s. The straighter black curve removes the seasonal variations.

Source: National Oceanographic and Atmospheric Administration (NOAA). The data are updated monthly at gml.noaa.gov/ccgg/trends.

tion has now surpassed 420 parts per million, which is only a little more than 0.04%.[10] In other words, carbon dioxide represents only a tiny fraction of the molecules in Earth's atmosphere. Nevertheless, as we've already discussed, this small amount of carbon dioxide is very important because of its role in the greenhouse effect.

Q What are the small wiggles on the graph?

The small wiggles represent seasonal variations in the carbon dioxide concentration. They occur because land-based plants and trees absorb carbon dioxide as they grow in spring and summer, then release it as they decay in fall and winter. The global pattern follows the seasons of the Northern Hemisphere because, as you'll see if you look at a globe, that is where most of Earth's land mass — and hence the land-based plants and trees — is located. (The Southern Hemisphere is mostly ocean.) The seasonal wiggles peak each year in May because that is when most of the prior year's vegetation has decayed (releasing its carbon dioxide) and the Northern Hemisphere's summer growing season is not yet far enough along for new vegetation to have absorbed very much carbon dioxide.

---

10 Here's how to see this: 400 parts per million would mean a concentration of 400 ÷ 1,000,000, which equals 4/10,000. Percentages are essentially parts per 100, making 4/10,000 the same as 0.04/100, or 0.04%. So a level of 420 parts per million is a bit higher than 0.04%.

Q Are measurements at Mauna Loa really representative of the whole world?

The carbon dioxide concentration varies somewhat from place to place on Earth, so it's important to choose measurement locations that are relatively unaffected by local conditions and therefore representative of changes occurring in the atmosphere as a whole. The Mauna Loa site was selected in the 1950s by Scripps Institution scientist Charles David Keeling because its high altitude and its location on a relatively isolated island make the air above it representative of a large portion of the global atmosphere. Today, scientists also measure the carbon dioxide concentration at many other locations around Earth, and these measurements confirm that the concentration is rising at a similar rate around the world. Note also that measurements are made by several independent scientific groups, and these different measurement sets agree very well, further confirming that the measured changes are real. We usually show the Mauna Loa data because they constitute the longest continuous record from any site and because other data confirm that they are representative of the global carbon dioxide concentration. Incidentally, Keeling's work has proven to be so important that the graph in figure 1.8 is often called the *Keeling curve* in his honor.

Q Can we measure the carbon dioxide concentration further into the past?

Although we have direct measurements only since the 1950s, scientists have discovered a variety of ways to estimate and measure the carbon dioxide concentration from earlier times.[11] The most reliable records come from air bubbles trapped in ancient ice — that is, ice that has remained frozen for long periods of time in glaciers or in the Greenland or Antarctic ice sheets. Although the work is difficult and requires great care, the basic idea is simple. An ice sheet represents the accumulated snow of many years, compressed over time into solid ice. Scientists sample this ice by drilling down to bring up an *ice core* (figure 1.9), which is marked by annual layers that arise from differences (in size and other characteristics) between summer and winter snow crystals. These layers can be used much like tree rings to determine the age of the ice at different depths, and air bubbles trapped within the layers represent samples of the atmosphere from those past times.

The longest ice core drilled to date extended to a depth of about 3.2 kilometers (2 miles) in the Antarctic ice, and it represents snows that fell and accumulated over a period of 800,000 years.[12] By studying air

---

11 FYI, because you are likely to hear the term when reading elsewhere about climate: Scientists often refer to any type of data that indirectly tell us about the past as a "proxy," a word also used in other fields (such as law and computer science) to mean "presumed to represent someone or something else." For example, ice core data are often said to provide a proxy for past carbon dioxide or temperature levels.

12 As this book goes to press in 2023, the European "Beyond EPICA" project (the original EPICA project obtained the 800,000-year ice core) has begun drilling with the aim of collecting ice samples dating back to 1.5 million years ago by about 2026; several other efforts to obtain very old ice, including the U.S. COLDEX project, are also in various stages of development.

Figure 1.9. The five photos on the left show the drilling of an ice core, which consists of compressed snow (with trapped air bubbles) laid down year after year; the long ice core is cut into sections for transport, storage, and study. The photo on the right shows the layering in a 1-meter-long section of an ice core; this particular section represents snow that fell about 16,250 years ago.

Source: (left) NASA, Goddard Space Flight Center; photos copyright by photographer Reto Stöckli, reprinted with permission; (right) USGS, National Science Foundation Ice Core Facility.

bubbles in this long ice core, scientists have been able to construct an 800,000-year record of Earth's atmospheric carbon dioxide concentration. Figure 1.10 shows the results, along with a zoom-out of the direct measurements made since the 1950s. Notice that the carbon dioxide concentration has risen and fallen substantially many times over the 800,000-year period. These variations must be natural, because they predate the human burning of fossil fuels, which became important only in the past couple hundred years.

Several key facts should jump out at you as you study figure 1.10:

- The carbon dioxide concentration has varied naturally over the past 800,000 years, but only within a range between about 180 and 290 parts per million. These natural variations are dwarfed by the huge increase that has occurred since the industrial revolution began in about 1750.

- Today's carbon dioxide concentration of about 420 parts per million is about 50% higher than the preindustrial concentration (280 parts

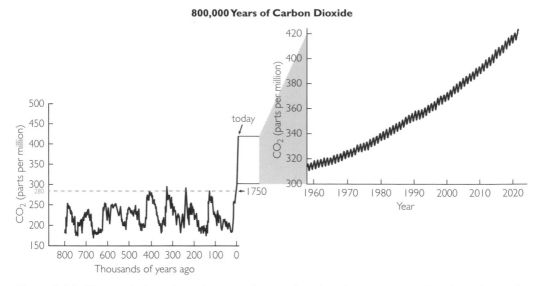

Figure 1.10. This graph shows how the atmospheric carbon dioxide concentration has changed over the past 800,000 years. The inset zooms in on the time period covered by direct atmospheric measurements.

Source: Data from the European Project for Ice Coring in Antarctica (EPICA). Also see the animated version of this graphic at gml.noaa.gov/ccgg/trends/history.html.

per million in 1750), which was itself near the highest found at any time in the past 800,000 years.

- If we extrapolate into the future (and assume that we continue to release carbon dioxide at recent rates), we find that the carbon dioxide concentration is on track to reach *double* its preindustrial value (560 parts per million) in only about 50 more years and *triple* that value (840 parts per million) by the middle of the next century.

Q How do we know the added carbon dioxide is a result of human activity, rather than natural sources?

It's true that there are natural sources (such as volcanoes) that can add carbon dioxide to the atmosphere, but we can be very sure that the recent dramatic rise in carbon dioxide that you see in figure 1.10 is due almost entirely to human activity. Most of this carbon dioxide comes from the burning of fossil fuels, with lesser amounts from deforestation and industrial processes, such as cement production,[13] that release carbon dioxide. Three major lines of evidence explain how we can be

---

13 Cement production involves the heating of carbonate minerals (such as limestone), which then release some of their stored carbon dioxide. The cement wedge in figure 1.11 shows this aspect of cement production, which currently contributes about 5% of the carbon dioxide that we are adding to the atmosphere. (Cement production also requires energy, which currently comes primarily from the burning of fossil fuels, but this part is counted in the fossil fuel wedges of figure 1.11.) Numerous companies are working on new cement production methods that offer the promise of reducing or eliminating these carbon dioxide emissions.

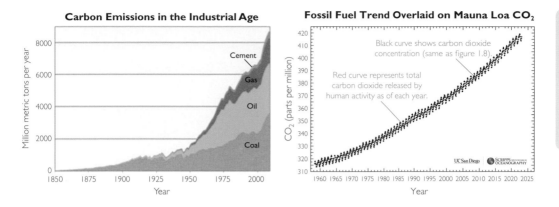

Figure 1.11. Left: This graph shows the annual amounts and sources of carbon dioxide released by human activity since 1850. Right: This graph demonstrates that the rate of increase in the total amount of carbon dioxide released by human activity (red curve) tracks almost perfectly with the measured rise in the atmospheric carbon dioxide concentration (black curve).

Sources: (left) Reproduced directly from J. M. Melillo, T. C. Richmond, and G. W. Yohe, eds., 2014, Highlights of Climate Change Impacts in the United States: The Third National Climate Assessment (U.S. Global Change Research Program); (right) Scripps CO2 Program (scrippsco2.ucsd.edu/graphics_gallery/mauna_loa_record/mauna_loa_fossil_fuel_trend).

sure that humans are responsible for the current rise in the carbon dioxide concentration.

First, the rise in atmospheric carbon dioxide coincides almost perfectly with the increased release of carbon dioxide by human activity. The ice core data tell us that for the 1,000 years prior to 1750, the atmospheric carbon dioxide concentration stayed very close to 280 parts per million. The dramatic rise that has since occurred began right about when the industrial revolution began, which is also when humans first began to use large quantities of fossil fuels. The correlation becomes even clearer when you look at how the release of carbon dioxide due to human activity tracks with the increase in the carbon dioxide concentration. The left-hand graph in figure 1.11 shows how the amount of carbon dioxide released by human activity each year has been rising with time and identifies the amounts from different sources. Now look at the right-hand graph in figure 1.11. At first, you may think you are just looking at a repeat of figure 1.8 with different colors, and the carbon dioxide data with its seasonal wiggles (now shown in black) is indeed the same. However, in this new figure, the red curve running up the center represents the rising total (cumulative) amount of carbon dioxide that is in the atmosphere due to human activity.[14] The virtually

14 To be a little more precise, the red curve in the right-hand graph in figure 1.11 represents the cumulative emissions adjusted by the fraction that remains in the atmosphere (as opposed to being absorbed by the land or oceans). The units for the red curve are not shown, because the important fact is the match in the *rate* of rise of the two curves. For more details on the graph, see the website of the Scripps CO_2 Program (scrippsco2.ucsd.edu).

perfect tracking leaves little room for doubt about our responsibility for the carbon dioxide increase. Moreover, there really isn't any other possibility. Scientists have a variety of ways to measure the amount of carbon dioxide that is added by natural sources, such as volcanoes, and it just doesn't compare to the amounts being released by the burning of fossil fuels and other human activity. In fact, the natural contributions to the rising carbon dioxide concentration are smaller than about 1% of the human contributions.

Q Wait — I heard that the amount of carbon dioxide released by human activity is very small compared to the amount released by the oceans and living organisms. So why do you say that natural sources aren't contributing to the increase?

It's true that the oceans and living organisms release large quantities of carbon dioxide, but these natural sources generally absorb just as much carbon dioxide as they release, creating a natural balance that therefore has no net effect on the atmospheric concentration of carbon dioxide. For example, not counting the extra added by human activity, the oceans always absorb essentially the exact same amount of carbon dioxide as they release, and plants naturally absorb all the carbon dioxide exhaled by animals (and people). The only way that natural processes can change the carbon dioxide concentration is by being out of balance. Volcanoes can disrupt the balance, since their eruptions add carbon dioxide without removing it, but as noted above, these imbalanced amounts are small compared to the human release. Deforestation also adds carbon dioxide, because it releases carbon dioxide stored in trees and plants — but this isn't exactly a natural process, since humans are the cause of most of the deforestation that has occurred in the past few centuries.

Q Does all the carbon dioxide released by human activity add to the atmospheric concentration?

No. Careful measurements show that only about half of the carbon dioxide released by humans each year stays in the atmosphere. Much of the rest is dissolved into the oceans (and some is taken up by plants and soil on land). Measurements confirm that the carbon dioxide concentration is increasing in the oceans in tandem with the increase in the atmosphere. This increase is making ocean water slightly more acidic, which creates the problem of *ocean acidification* that we will discuss in chapter 3.

Q How do the "million metric tons" on the left side of figure 1.11 relate to "parts per million" on the right?

Although the atmospheric concentration of carbon dioxide is measured in parts per million (ppm), human emissions are usually measured by their mass in metric tons. (1 metric ton = 1,000 kilograms, which is about 2,200 pounds or 1.1 U.S. tons.) There's a subtlety, however: Some sources tend to give the mass of the carbon alone (not counting the mass of the oxygen in the carbon dioxide), while others tend to give the full mass of the carbon dioxide. Fortunately, it's easy to convert between the two: To convert from carbon mass to carbon dioxide mass, simply multiply by 44/12 (because CO_2 has a molecular weight of 44 while carbon has an atomic weight of 12); divide to go the other direction. As for the relationship between metric tons and ppm, there's a simple equivalence based on the total mass of carbon dioxide in Earth's atmosphere: Each single ppm of carbon dioxide represents a mass of 2.13 billion tons of carbon, which is the same as 7.8 billion tons of carbon dioxide.

Second, as is the case with any fire, the burning of fossil fuels consumes oxygen at the same time that it releases carbon dioxide, which means that if the rising carbon dioxide concentration comes from fossil fuels, there should be a corresponding *decrease* in the atmospheric concentration of oxygen. The expected decrease has been measured (figure 1.12) and its rate matches what we expect if it is being caused by the combustion (burning) of fossil fuels.

The third and most convincing line of evidence comes from careful chemical analysis of atmospheric carbon dioxide, which shows changes that make sense only if fossil fuels are the source of the rising concentration. The key to understanding this evidence is to know that carbon atoms come in three different forms, or *isotopes*, known as carbon-12, carbon-13, and carbon-14, and the relative abundances of these three isotopes are different in carbon that comes from different sources (such as volcanoes, deforestation, and the burning of fossil fuels). Therefore, we can determine the source of atmospheric carbon dioxide by measuring these isotope abundances.

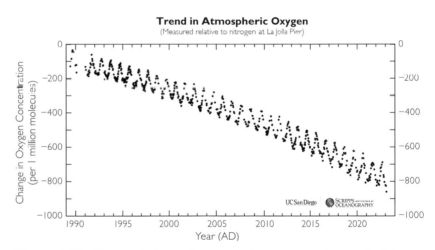

Figure 1.12. This graph shows that oxygen in our atmosphere is declining as expected if we assume that the cause is the combustion of fossil fuels. Note that the total loss (since the late 1980s) is about 800 per 1 million molecules, or 0.08%; we don't expect direct impacts from this loss because it is so small compared to the overall atmospheric oxygen concentration of about 21%.

Source: Scripps O2 Program (scrippso2.ucsd.edu).

Let's first consider carbon-14, which is radioactive and exists on Earth only as a result of ongoing production as cosmic rays from space hit atoms in Earth's upper atmosphere.[15] Carbon-14 becomes incorporated into living organisms through respiration, but it decays after the organisms die, and the organisms that made fossil fuels died so long ago that there is no carbon-14 in fossil fuels at all. As a result, if fossil fuels are the source of the rising carbon dioxide concentration, the relative abundance of atmospheric carbon-14 (compared to ordinary carbon-12) should be falling as the total carbon dioxide concentration rises — and this is just what has been observed.

Even more impressive evidence comes from changes in the relative abundance of carbon-13 (Figure 1.13). Overall, carbon-13 represents about 1.07% of all natural carbon on Earth, but its percentage is slightly lower in living organisms (because life incorporates carbon-12 into liv-

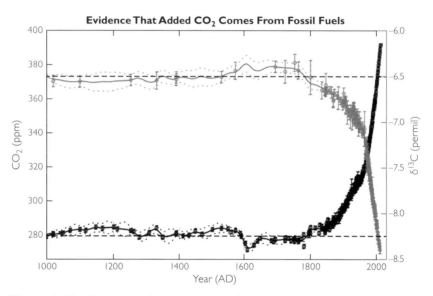

Figure 1.13. This graph shows the atmospheric carbon dioxide concentration (black) and the relative abundance of carbon-13 (brown) over about the past 1,000 years, as measured from ice cores. The declining abundance of carbon-13 is a "smoking gun" that leaves no doubt that the rising carbon dioxide concentration is coming mostly from the burning of fossil fuels.

Note: The carbon-13 abundance is given in terms of "$\delta^{13}C$," which is a standard scientific unit used for this purpose. Source: M. Rubino et al., J. Geophys. Res. Atmos. 118 (2013): 8482–8499, doi:10.1002/jgrd.50668.

---

15 "Radioactive" materials have atomic nuclei that tend to undergo *decay*, meaning they change into another form over time; the time is characterized by the *half-life*, which represents the time it takes for half of the nuclei in any sample to decay. Carbon-14 (which decays to become nitrogen-14) has a half-life of about 5,700 years, which means that any carbon-14 that existed when Earth first formed is long gone. The fact that carbon-14 decays with a known half-life in the remains of living organisms that have died is what makes *radiocarbon dating* possible for fossils and archeological artifacts up to a few tens of thousands of years old. (In older fossils, the carbon-14 will have all decayed, so scientists determine their ages from radioactive isotopes that have longer half-lives.)

ing tissues more readily than carbon-13), which means it is also slightly lower in fossil fuels (since they are the remains of living organisms). Figure 1.13 shows ice core data for the past 1,000 years, with the total carbon dioxide concentration in black and the relative abundance of carbon-13 in brown. Notice that the carbon-13 abundance has been dropping in tandem with the rise in carbon dioxide, just as we expect if the rise in carbon dioxide comes from the burning of fossil fuels (with their lower abundance of carbon-13). In effect, these isotopic data are a "smoking gun" that leaves no doubt that most of the added carbon dioxide is coming from the burning of fossil fuels.

Is human activity also increasing the concentrations of other greenhouse gases?

Yes. The left-hand graph in figure 1.14 shows changes in the atmospheric methane concentration since the late 1970s (the time period for which direct measurements of all the gases shown in the figure are available). Data from ice cores and other sources indicate that the methane concentration has more than doubled since 1750. Human activity adds methane to the atmosphere in several ways, but the three largest are (1) agriculture, in which methane is released from sources that include rice paddies and livestock; (2) oil and gas extraction and transport, during which methane can leak directly into the atmosphere; and (3) landfills, in which decomposing waste releases methane.

The middle graph in figure 1.14 shows the rapidly rising nitrous oxide (N_2O) concentration. Nitrous oxide is released primarily through the production and use of fertilizers, which means these emissions are tied to agriculture.

The rightmost graph in figure 1.14 shows halocarbons, such as CFCs, which come entirely from human manufacturing and do not exist naturally. Notice that while CFC concentrations were rising rapidly in the 1970s and 1980s, their concentrations have since declined. The reason for the decline is that, starting in the 1970s, scientists began to recognize that CFCs were causing destruction of Earth's atmospheric ozone

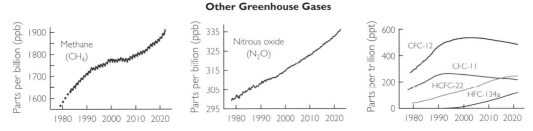

Figure 1.14. These graphs plot the concentrations, measured since the late 1970s, of methane, nitrous oxide, and halocarbons, including CFCs and gases used as substitutes for CFCs.

Source: NOAA Global Monitoring Laboratory (gml.noaa.gov/aggi/).

layer, which protects us from dangerous ultraviolet light from the Sun. As a result, the nations of the world came together to sign the global treaty known as the Montreal Protocol (and subsequent revisions to strengthen it), which successfully led to a dramatic decline in the production and use of CFCs.

To gauge the combined effects of carbon dioxide and other greenhouse gases, scientists use something called the *annual greenhouse gas index* (AGGI), which you can learn more about from the AGGI section of the NOAA website (gml.noaa.gov/aggi). Measurements of the AGGI indicate that the combined greenhouse warming from *all* gases emitted through human activity each year is roughly 50% larger than that from carbon dioxide alone.

It's worth noting that, with the exception of the methane leaks from oil and gas extraction and transport, most of these additional emissions come from sources that are distinct from the use of fossil fuels. They therefore represent separate problems that we will ultimately need to deal with along with the primary issue of carbon dioxide emissions. More specifically, since CFCs are declining already, the most important "other" greenhouse gas problems are agricultural emissions (methane and nitrous oxide) and landfill emissions (methane). I don't mean to downplay the importance of these issues (or of carbon dioxide from cement production, which is also distinct from fossil fuels), but I'll continue to focus on energy-related carbon dioxide in this book for two major reasons. First, carbon dioxide emissions are by far the largest contributor to global warming. Second, these "other" greenhouse gases do not remain in the atmosphere for nearly as long as carbon dioxide (typically about a decade for methane and a century for nitrous oxide, compared to millennia for carbon dioxide), which means that once we find ways to limit their emissions, they'll return to their natural levels relatively quickly.

Q Why do I sometimes hear that methane and nitrous oxide are more "potent" greenhouse gases than carbon dioxide?

The "potency" of a greenhouse gas describes how much energy it can trap per unit of mass. More specifically, scientists define the *global warming potential* (GWP) of a gas in terms of how much energy one ton of the gas can trap over some period of time compared to one ton of carbon dioxide. For example, the global warming potential of methane over a century is approximately 28, meaning that a ton of methane leads to 28 times as much warming over a century as a ton of carbon dioxide.[16] The global warming potentials of nitrous oxide and of CFCs are even higher. These high global warming potentials explain how the very low concentrations of these gases (notice in figure 1.14 that they are measured in parts per *billion*) can nev-

---

16 The global warming potential can vary significantly depending on the time period considered; for example, methane's global warming potential is above 80 for a time period of 20 years.

ertheless make the overall greenhouse effect significantly stronger than it would be from carbon dioxide alone.

Q What's the bottom line for Fact 2 (human activity is adding carbon dioxide)?

We are left with no doubt that human activity is adding carbon dioxide and other greenhouse gases to the atmosphere. The rising concentration has been carefully measured and documented, and we discussed three distinct lines of evidence proving that the carbon dioxide rise is due to human activity. Moreover, as Margaret Thatcher said in the quote that opens the chapter, we are adding carbon dioxide at an "unprecedented rate" that is "new in the experience of the Earth," at least for the past 800,000 years (and probably much longer).

Data Supporting the Inevitable Conclusion

Statement 3 of our global warming 1-2-3 — that we should *expect* global warming to be occurring — follows logically from Facts 1 and 2. However, in science, even seemingly unassailable logic should always be confirmed. So let's look briefly at the evidence confirming that a warming is under way.

Q What data show that the world is actually warming up?

To determine whether Earth is actually warming up, scientists need to track changes in Earth's global average temperature over time. Direct measurements from which we can infer global temperature go back to about 1880, though there are greater uncertainties for the earlier years.

Figure 1.15 shows the data. Notice the clear upward trend, with an overall gain of about 1.1°C, or 2°F, over the past century. This confirms that our world is warming, just as our 1-2-3 logic tells us to expect.

In addition to the fairly obvious general trend, it's worth noting at least two additional facts visible in figure 1.15:

- If you look closely, you'll see that 1998 was unusually hot compared to the years on either side of it (for reasons relating to El Niño, which we'll discuss starting on page 38). But if we ignore that one year, *every single year in this 21st century has been hotter than any year in the 20th century.*
- The most recent nine years shown (through 2022, which was the most recent year for which data were available when this book went to press) represent the nine hottest years on record.[17]

---

17   This statement is based on the NOAA temperature record shown in figure 1.15. Other data sets (see figure 1.16) would make the statement true for the last eight (rather than nine) years.

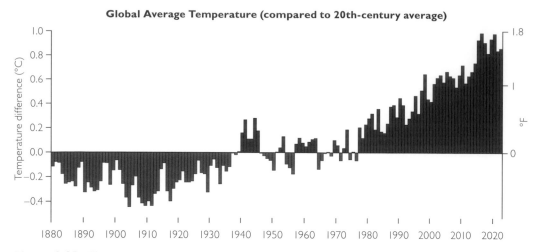

Figure 1.15. This graph shows how global average temperature varied from 1880 through 2022. The horizontal line (a difference of 0°) represents the average temperature for the entire 20th century. Notice the clear warming trend of recent decades.

Source: National Climate Data Center (NOAA).

Q Are the temperature data reliable?

Yes, but only thanks to some very careful work. Measuring Earth's global average temperature essentially requires scientists to average local temperature measurements from many places around Earth, and this is not easy to do. There are three fairly obvious complexities. First, even today, there are large regions of our planet (including the oceans and regions near the poles) for which we have relatively few temperature measurements, making it difficult to come up with a fully global average. Second, this problem becomes worse as we look to the past, when there were fewer weather stations. Third, many measurements are made in or near urban areas, which tend to produce higher temperatures than they would if they were rural or unpopulated, because urban areas generate their own heat (the "urban heat island effect") through such means as the absorption of sunlight by pavement and the emission of heat from cars and homes.

Because of these and other difficulties, there's always some uncertainty in Earth's precise global average temperature. In fact, the estimate of 15°C that I stated earlier (see figure 1.1) could be off by as much as a degree or two. That is why figure 1.15 shows only temperature *differences* (scientists often call them "anomalies") from year to year, rather than actual values. To understand how this helps, imagine weighing yourself every day on two different scales, one of which always gives you a lower weight than the other. You may not have any way to know which scale is showing your true weight, but if you actually lose two pounds in a week, both scales will probably show the same two-pound loss. In much the same way, year-to-year differences measured by weather stations are much more reliable than their exact temperature readings. Therefore, by averaging year-to-year differences measured at weather stations around the world, scientists can get a reliable record of how Earth's temperature is changing, even without knowing the "true"

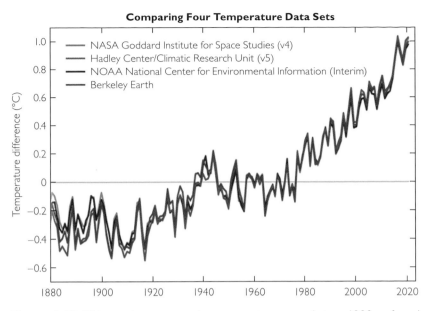

Figure 1.16. This graph compares the temperature record since 1880 as found (based on available data records) by four different research groups, each essentially independently of the others. The close agreement among all four groups gives us great confidence that the measured warming trend is irrefutable.

Note: The zero line here is the average for the period 1951–1980, which is slightly different from the 20th-century average used as the zero line in Figure 1.15. Source: Compiled by Gavin Schmidt, NASA, from the four data sets.

average temperature. Moreover, for recent decades, scientists also have data from satellites,[18] which in effect can take measurements from all around the world, including the regions where no weather stations are located.

That said, it's still not easy. For example, the numbers and locations of weather stations change over time, the heat in cities can change as they grow, and different satellites collect data in different ways. Scientists must be very careful to take these factors into account when computing the change in temperature from one year to the next. Fortunately, several different scientific groups analyze both ground and satellite temperature data, each using somewhat different techniques.[19] The results found by these different groups are all in close agreement (figure 1.16), giving scientists great confidence that the trend shown in figure 1.15 is real. Indeed, while there is some debate over the size of the uncertainties in the data, there is no serious controversy over the general trend, which clearly shows that the world has been getting warmer over the past century.

---

18 However, satellites cannot directly measure temperature at the surface, instead giving readings for temperatures at roughly the altitudes at which airplanes fly (8–15 kilometers), where warming is less pronounced than at the surface. For this reason, satellite temperature measurements must be interpreted with great care.

19 A good starting point for learning more about how these groups measure the global average temperature is carbonbrief.org/explainer-how-do-scientists-measure-global-temperature/.

Q **How much uncertainty is there in the temperature data in figure 1.15?**

Scientists generally state measurement uncertainty (often referred to as the "margin of error") in terms of some level of confidence, and the level most commonly used is "95% confidence" (which you may hear scientists refer to as "two sigma"). For example, a measurement stated as "1.1°C with an uncertainty of 0.1°C" (often written as 1.1°C ± 0.1°C) implies a 95% chance that the true value is between 1.0°C and 1.2°C. Note that these types of uncertainty statements are *not* mere guesswork; they are based on careful analysis of the data and of potential sources of error. As you might expect, the uncertainties in the temperature data are greater for times further in the past (when there were fewer weather stations and no satellite measurements). Overall, they are approximately as follows:[20]

- For the early years in figure 1.15 (e.g., 1880–1900), the uncertainty in the measurements (with 95% confidence) is about 0.1°C. For example, the bar for 1885 shows –0.2°C, so the true value (with 95% confidence) was likely between –0.3°C and –0.1°C.

- The uncertainty becomes smaller as time goes on, and for recent decades (since about 1980) it is down to about 0.03°C. For example, the bar for 2015 shows a value of 0.90°C, so the true value (with 95% confidence) was likely between 0.87°C and 0.93°C.

- For the overall warming trend of 1.1°C since 1880, the uncertainty is about 0.2°C, so the total warming (with 95% confidence) has probably been between 0.9°C and 1.3°C.[21]

Q **Are "aerosols" masking some of the warming?**

Aerosols are tiny particles suspended in the atmosphere. Aerosols are relevant to global warming because they reflect sunlight and therefore have a cooling effect, so more aerosols means less warming. Most aerosols are natural (coming from sources such as volcanoes, forest fires, sea salt in ocean spray, and blown dust), but human activity adds aerosols through air pollution and human-caused fires. This means that as air pollution has worsened and fires have increased around the globe, the released aerosols have been countering some of the effects we'd otherwise be seeing from the rising greenhouse gas concentration. Estimates suggest that Earth would already have warmed by up to another few tenths of a degree if not for these pollution-related aerosol effects. So bottom line: Yes, aerosols have likely been masking some of the warming, which means that if we cleaned up air pollution and stopped burning so many fires — both of which are generally good ideas for health reasons — we might actually make global warming worse unless we reduce greenhouse emissions at the same time.[22]

20 You can find further discussion of the uncertainties at this NASA website: data.giss.nasa.gov/gistemp/. For a deeper dive into the statistical analysis, see J. Hansen et al., "Global Surface Temperature Change," *Rev. Geophys.* 48, RG4004 (2010).

21 Using a somewhat more sophisticated analysis, the Sixth Assessment Report from the Intergovernmental Panel on Climate Change (IPCC), released in 2021, concluded that human activity has caused a total warming to date of between 0.8°C and 1.3°C, with 1.1°C the most likely value.

22 A great starting point for learning more about aerosols and their effects is the following NASA website: earthobservatory.nasa.gov/features/Aerosols.

Science Summary: Global Warming 1-2-3

Let's repeat the "global warming 1-2-3" that we started out with:

1. ***Fact:*** Carbon dioxide is a greenhouse gas, by which we mean a gas that traps heat and makes a planet (such as Earth or Venus) warmer than it would be otherwise.

2. ***Fact:*** Human activity, especially the use of fossil fuels — by which we mean coal, oil, and natural gas, all of which release carbon dioxide when burned — is adding significantly more of this heat-trapping gas to Earth's atmosphere.

3. ***Inevitable Conclusion:*** Given that more carbon dioxide means warmer temperatures and that we are adding carbon dioxide (and other greenhouse gases) to Earth's atmosphere, it is inevitable that global warming should occur as a result. The more of this gas we add, the greater the warming will be.

In this chapter, I've shown you the evidence demonstrating that Facts 1 and 2 are scientifically supported beyond any reasonable doubt, along with data confirming that the inevitable warming is already under way. I hope that you now see that, despite the complexity of Earth's overall system, global warming itself is easy to understand.

2 The Skeptic Debate

Twenty-five years ago people could be excused for not knowing much, or doing much, about climate change. Today we have no excuse.

— Archbishop and Nobel Peace Prize Laureate Desmond Tutu, Apr. 10, 2014 (in an op-ed he wrote for *The Guardian*)

If we are going to create a pathway to a post–global warming future, we will need broad public agreement about the reality of the problem we wish to solve. Unfortunately, this agreement has often been hard to come by, in large part because of a relatively small number of "skeptics" who have disputed the reality and/or seriousness of global warming. For this reason, we'll turn our attention in this chapter to the most common skeptic claims. To keep the discussion streamlined, I'll focus on four major points of debate that skeptics have raised over the years. In each case, we'll look at the evidence that makes most scientists reject the skeptic claims, and we'll also use the opportunity to discuss climate science in a little more detail.

It's important to note that, at least among those skeptics who have relevant scientific credentials, none of the dispute concerns the basic 1-2-3 science behind global warming. Instead, it concerns questions of how rapidly our planet is warming, whether factors besides human activity are involved, and the magnitude of the threat that the warming poses.

In other words, the skeptic debate is ultimately a matter of risk assessment. Most scientists have concluded that the threat posed by global warming is significant and demands immediate action, while the skeptics argue that the threat is being overblown. So as you read about the skeptic claims, I suggest keeping the following question in mind: Given that we *know* from the basic science that an increasing carbon dioxide concentration causes global warming, how much risk are you willing to accept?

Skeptic Claim 1: It's Not (or No Longer) Warming Up

The first type of skeptic claim we'll consider goes to the question of whether Earth is actually undergoing the warming predicted by the 1-2-3 science. In essence, this type of claim suggests that there might be something — such as a self-regulating (or negative) feedback process in the climate system — that might prevent or slow the expected warming. This type of claim has become much less common in recent years, largely because the warming trend (see figure 1.15) has been increasingly difficult to dismiss. It still crops up on occasion, however, so let's take a brief look at major questions that skeptics have raised about the observed warming trend.

Q Could the data in figure 1.15 be wrong?

The earliest claims of this type questioned the validity of temperature measurements like those shown in figure 1.15. You may still sometimes hear these measurements questioned in the media, but the fact that four independent groups have tracked the temperature data and all come up with very similar results (see figure 1.16) makes it exceedingly unlikely that we could be misinterpreting the general trend.

It's also worth noting that one of the people best known for questioning the temperature data — a physicist by the name of Richard Muller — was so concerned about the data's accuracy that he decided to check it himself. In 2010, with much of his funding coming from groups that opposed climate action, he founded what is now called Berkeley Earth to challenge and check the data. After some two years of work, he and his team announced their results: They found no significant accuracy issues and concluded that the temperature trend is real and human activity is the cause of the warming.[1] While this conclusion was no surprise to the vast majority of climate scientists, the fact that it came from a former skeptic largely shut down the debate over the accuracy of the data. Berkeley Earth is now one of the major groups tracking global temperature changes (see figure 1.16).

Q Did global warming stop after the late 1990s?

If you go back to media reports prior to about 2016, you'll find that even though few skeptics still questioned the basic data, many were claiming that global warming had "stopped" (or "paused") after the

---

1 For more detail, read Muller's op-ed "The Conversion of a Climate-Change Skeptic" in *The New York Times*, July 28, 2012.

late 1990s. Today, one look back at figure 1.15 belies this claim, but you can see its origin by putting your thumb over the data from 2015 onward; the upward trend since the 1990s then becomes much more difficult to see. But even at that time, there were two simple ways to show that the world was in fact still warming. The first was to recognize that the climate system has some natural variability, which means that long-term trends are better revealed by multiple-year averages than by data for single years. In figure 2.1, the year-by-year data from figure 1.15 have been replaced with the *average* (mean) for each five-year period. Notice that every five-year period since 1980 has set a new record for the hottest (since 1880), and while there was a *slowing* of the upward trend from the late 1990s to 2015, the warming certainly did not stop.

The second way to see that the warming hadn't stopped requires thinking a little more generally about the greenhouse effect. Recall that the greenhouse effect acts like a blanket, trapping energy near Earth's surface (see figure 1.4), so a strengthening greenhouse effect means more of this trapped energy. This "extra" energy can appear in the climate system in many different ways, and a rising surface temperature is only one of them. In fact, more than 90% of the added energy is expected to warm the water *in* the oceans (as opposed to warming the land and ocean surface). Figure 2.2 shows ocean temperature data for as far back as we have good measurements — and as you can see, there's no hint of any pause in the warming trend since the 1980s.

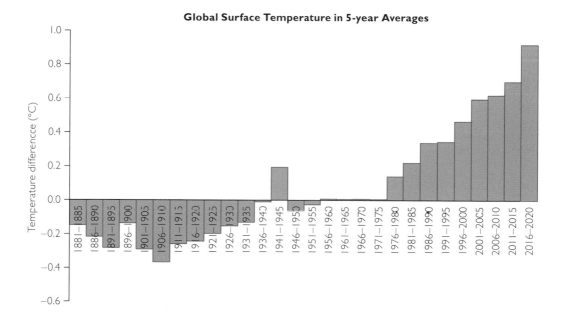

Figure 2.1. This graph shows the temperature data from figure 1.15 grouped into five-year averages (through 2020). Notice that every five-year period since 1980 has set a new record.

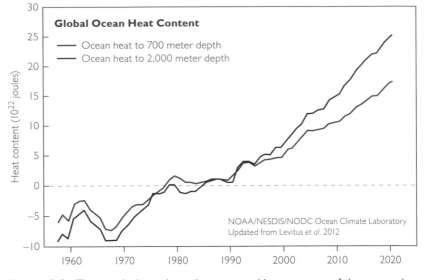

Figure 2.2. This graph shows how the measured heat content of the oceans has changed in recent decades. The data are plotted as five-year moving averages, meaning the average from two years before to two years after each individual year (for example, the 2020 value represents the average for 2018–2022). Notice that there has been no general slowing of the rise in ocean heat content, and in recent years more of the heat has been appearing in deeper waters.

Source: NOAA; most of the data come from the Argo program, which makes measurements with about 4,000 floats dispersed around the oceans (see argo.ucsd.edu).

Q Why is there so much year-to-year variability in the surface temperature trend (as shown in figure 1.15)?

The answer lies largely in the fact that more than 90% of the added energy is going into the oceans, which means that small changes in the ocean heat uptake can have significant effects on surface temperatures. Some years might have a little more of the "extra" energy due to the strengthening greenhouse effect going into surface warming and a little less into the oceans, and other years might have the reverse.

Digging a little deeper, you might wonder what types of climate processes explain why the heat would be deposited in different ways at different times, and this brings us to a climate phenomenon known as El Niño. El Niño events typically last for about a year (the precise length varies) and recur at irregular intervals (typically between two and seven years). They are usually recognized through a warming of the central and eastern Pacific Ocean[2] (figure 2.3), but that is only the beginning of their climate impacts. Other major effects of El Niño include a weakening of the normal east-to-west equatorial winds (the "trade winds") and shifts from the normal posi-

---

2 The name El Niño comes from the eastern Pacific warming, which was first identified by South American fishermen in the 1600s. Because these fishermen noticed that the warming often peaked around Christmas time, they referred to it as El Niño de Navidad (Spanish for "the Christ child"), which was later shortened to El Niño (Spanish for "the little boy").

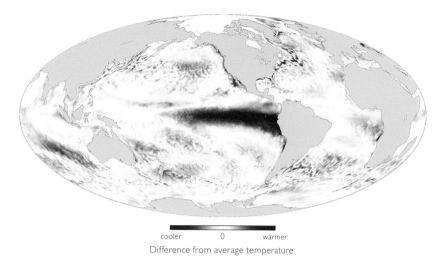

cooler   0   warmer

Difference from average temperature

Figure 2.3. This map shows a typical example of the changes in ocean surface temperature that occur during an El Niño event. Notice the warming of the central and eastern tropical Pacific.

Source: NOAA Climate.gov (www.climate.gov/news-features/blogs/enso/what-el-niño–southern-oscillation-enso-nutshell).

tions of the high-altitude jet streams, both of which affect regional weather. In the United States, for example, El Niño tends to mean warmer and drier conditions in the northwest and wetter conditions in the south and southeast. Globally, years with El Niño events tend to be warmer than they would be otherwise, and the years that look warmest (compared to neighboring years) in figure 1.15 — such as 1998 and 2015/2016 — are those in which El Niño was especially strong.

El Niño is actually part of a larger, irregular climate cycle called the El Niño–Southern Oscillation (ENSO). The ENSO cycle encompasses three general types of condition: (1) El Niño, marked by a warmer central and eastern tropical Pacific; (2) La Nina ("the little girl"), which is essentially the opposite; and (3) neutral, which refers to times in between El Niño and La Niña events. You can often get a general sense of what the weather will look like in coming months by knowing where we are within the ENSO cycle. (Learn more at climate.gov/enso.)

Q Has global warming stopped more recently?

If you look again at figure 1.15, you'll notice that while the nine most recent years have been the nine warmest on record, the hottest single year (as of 2022) was 2016, which means it was followed by six years that were somewhat cooler. Perhaps you won't be surprised to learn that this has caused some skeptics to once again argue that "global warming has stopped." I won't spend too much time on this argument, because the data in figures 2.1 and 2.2 refute it in exactly the same way they refuted the earlier claim of this nature. So to answer the question: No, global warming did not stop after the 1990s, it did not stop after

2016, and it is highly unlikely that it will stop in the future unless we stop doing the things (primarily the burning of fossil fuels) that have been causing the warming to occur.

Q **Why was 2016 so hot?**

As discussed above, the answer is El Niño. The years 2015 and 2016 were marked by the strongest El Niño in decades — in fact, the strongest since 1998. This also means that while the subsequent six years were all cooler than 2016, a new hottest record is very likely when the next El Niño occurs — which may already have happened by the time you are reading this book.

Q **Is there any way that Skeptic Claim 1 ("not warming") could be correct?**

A couple of decades ago, there were still enough uncertainties in the temperature measurements that it might have been legitimate to raise questions about whether the warming trend was real. For that reason, scientists put a great deal of effort into understanding the uncertainties, and while some still exist, there is no longer any serious debate about the trend.

It's also worth remembering that the trend we are seeing matches what we should expect from the basic 1-2-3 science of global warming. Figure 2.4 repeats data from figure 1.15, but this time with an overlay showing the rising carbon dioxide concentration. Notice that the two

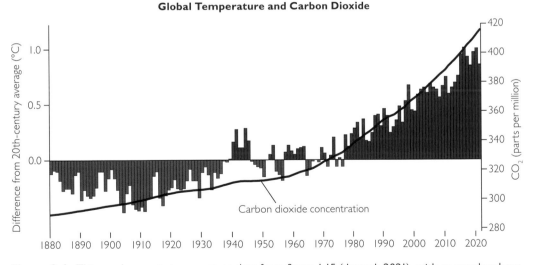

Figure 2.4. This graph repeats temperature data from figure 1.15 (through 2021), with an overlay showing the carbon dioxide concentration. Notice that the carbon dioxide concentration and the temperature have been rising in tandem for recent decades, just as we expect from the basic science of global warming.

Note: The CO2 data since the late 1950s are the direct measurements shown in figure 1.8 (an average for each year); earlier data are from ice cores. Source: NOAA climate.gov (based on work by Dr. Howard Diamond).

trends — the observed warming and the rising carbon dioxide concentration — are indeed moving hand in hand.

This brings us to the bottom line on this topic: There is simply no scientific doubt that the world is warming, and while there may be some ups and downs in the rate,[3] we can expect this warming to continue for as long as we continue to add greenhouse gases to Earth's atmosphere.

Skeptic Claim 2: It's Warming Up, but It's Natural

We now turn to the second type of skeptic claim, which accepts the fact that Earth is warming up but argues that the warming may be occurring for natural reasons, rather than as a result of human activity.

Q **Could the Sun be the cause of the observed warming?**

The Sun's energy output varies slightly (by much less than 1%) from year to year, leading some people to wonder if the Sun might be responsible for the recent warming trend. It's true that relatively small changes in the amount of sunlight reaching Earth can affect the climate; as we'll discuss shortly, such changes are thought to have been the triggers for cycles of past ice ages. However, we can be very confident that changes in the Sun's output are *not* the cause of the recent trend, because of the data shown in figure 2.5. This figure compares annual changes in Earth's temperature since 1880 (red curve) to changes in the amount of sunlight reaching Earth (blue curve). Notice that while the two trends matched up moderately well until about 1950, they have since gone in opposite directions. Clearly, we cannot blame an increase in temperature on a decrease in sunlight.

Q Can you explain the curves in figure 2.5 more clearly, including why they use an 11-year average?

Notice that each curve is actually two curves: a heavier one showing the 11-year average and a lighter one showing year-to-year data. Let's start with the red temperature curves. The light red curve represents the same data shown in figure 1.15, but as a line graph rather than a bar chart. The heavier red curve is what we call an 11-year *moving average* (or "running mean") drawn through the lighter curve, meaning that each value on the

---

3 Especially, for example, if there is a major volcanic eruption of the type that tends to occur at least once every few decades, which would have a temporary (up to a few years) cooling effect due to the aerosols added to the atmosphere.

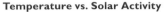

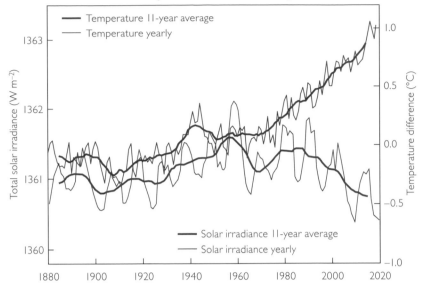

Figure 2.5. This graph compares changes in global average temperature from 1880 through 2020 (essentially the same data shown in figure 1.15) to changes in the Sun's output (as measured by its "irradiance," which is the average power of sunlight reaching Earth, shown in units of watts per square meter). Notice that, for recent decades, the Sun's output has moved in the opposite direction from the observed warming, which means the Sun cannot be the cause of recent global warming.

Source: NASA/JPL-Caltech; graph posted at climate.nasa.gov/causes.

heavier curve represents the average from five years before to five years after the year that the value represents (so a total of 11 years). (The heavier curve therefore stops five years before the end of the data set, since that is the last available 11-year average.) The blue curves are similar. The light blue curve shows actual year-to-year data, while the heavier blue curve shows an 11-year moving average. One subtlety: For recent decades, the solar irradiance data are based on actual measurements made by orbiting satellites, while earlier data are reconstructed based on historical observations of sunspots, which have been reliably recorded since long before the satellite era and correlate with irradiance quite well. This subtlety also explains why the graph uses 11-year averages. The number of sunspots on the Sun, and hence the solar irradiance, varies in an approximately 11-year cycle. Therefore, an 11-year moving average is the fairest way to show the data, because it effectively removes the variations due to the sunspot cycle so that we can see the underlying general trend.

That is already pretty definitive, but there's also a second major reason we can rule out the Sun as the cause of the recent warming. If the Sun were responsible for global warming, we would expect the extra sunlight reaching Earth to warm the surface and the entire atmosphere more or less uniformly. In contrast, while the greenhouse effect warms Earth's surface and lower atmosphere, it actually *cools* Earth's upper atmosphere (the stratosphere and above).[4] Observations confirm that the upper atmosphere is cooling, just as we'd expect with a strengthening greenhouse effect, but the opposite of what we'd expect if global warming were being caused by the Sun.

In fact, several additional patterns of warming are also consistent with a strengthening greenhouse effect but not consistent with changes in the Sun.[5] For example, only greenhouse warming can account for measurements showing that nights have warmed more than days and winters (in both hemispheres) have warmed more than summers. Moreover, satellite measurements indicate that the total heat radiating into space from Earth has declined at the specific wavelengths radiated by carbon dioxide, which can only mean that this heat is being trapped by carbon dioxide molecules through the greenhouse effect.

Q Is Mars also warming up?

In trying to point to the Sun as a culprit for climate change, some skeptics have claimed that Mars is also undergoing global warming — but there is no evidence to support this claim. Interestingly, this claim appears to have originated with a comparison of photos of Mars taken a couple decades apart (in 1977 and 1999). These photos did indeed suggest that the later time frame was warmer, but this could be fully explained by the fact that the earlier photo was taken shortly after one of Mars's famous global dust storms, during which dust in the atmosphere reflects sunlight and thereby makes the planet cooler. More careful evaluation of current evidence shows neither a general warming nor a general cooling trend during recent decades on Mars.

Q Could the warming be due to other natural factors besides the Sun?

Earth's climate is very complex and is affected by many factors, both human and natural, so it's worth exploring whether there might be any other natural process that might explain the observed warming. The primary way that scientists investigate this possibility is by using what we call *models* of the climate.

Scientific models differ from the models you may be familiar with in everyday life, which are typically miniature representations of real objects, such as model cars or airplanes. In contrast, a scientific model is a conceptual representation, often developed with the help of com-

---

4 The precise reasons why the greenhouse effect leads to upper atmospheric cooling are fairly complex and beyond the scope of this book, but they are based on detailed calculations of how the greenhouse effect works. If you want evidence that these calculations are valid, just look to Venus, where the extremely strong greenhouse effect causes not only the very high surface temperature that we've already discussed, but also a cool upper atmosphere that matches the predictions of greenhouse calculations.

5 For a more complete discussion of the "fingerprints" that indicate the warming is from the greenhouse effect and not the Sun or other natural factors, see skepticalscience.com/its-not-us.htm.

puters, that uses known scientific laws, logic, and mathematics in an attempt to describe how some aspect of nature works. The model can be tested by seeing how well it corresponds to reality. Models are important in almost every field of science, but here we'll focus specifically on models of Earth's climate.

The principle behind a climate model is relatively simple. Scientists create a computer program that represents the climate as a grid of cubes like those shown in figure 2.6, where each cube represents one small part of our planet over one range of altitudes in the atmosphere. The "initial conditions" for the model consist of a mathematical representation of the weather or climate within each cube at some moment in time. This representation might incorporate data on such things as the temperature, air pressure, wind speed and direction, and humidity at the time the model begins. The model uses equations of physics (for example, equations that describe how heat flows from one cube to neighboring cubes) to predict how the conditions in each cube

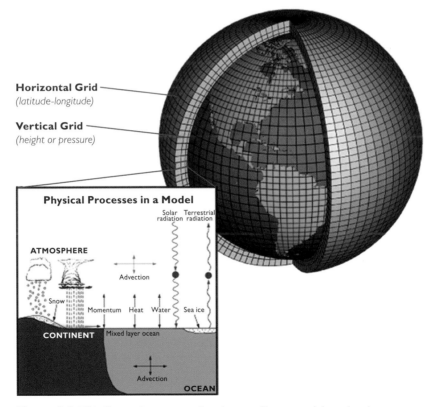

Figure 2.6. This illustration summarizes how a climate model works. A computer program represents Earth's climate through a series of cubes. In each cube, scientists input data from some point in time to represent "initial conditions" and then "run" the model, by using equations that represent physical processes, to see how the conditions change with time.

Source: NOAA.

will change over some time period, such as the next hour. It then uses the new conditions and the equations to predict the conditions after another hour, and so on. In this way, the model can simulate climate changes over any period of time.

Decades ago, climate models were fairly simple, using grids no more complex than the one in figure 2.6. Over time, however, scientists have in essence used trial and error (and more powerful supercomputers) to make the models better and better. Again, the principle is easy to understand: If your model fails to reproduce the real climate in some important way, you look to see what might be going wrong. For example, you might have neglected some important law of physics, or the cubes in your grid might need to be smaller to give accurate results. Once you think you know what went wrong, you revise the model. If the revised model works better, then you have at least some reason to think you are on the right track, and if it doesn't, you go back to the drawing board.

Q **Why are you saying "models" plural?**

It is not possible to create an exact representation of Earth's climate (because it is too complex), so approximations must inevitably be used. Over the past several decades, numerous research groups around the world have made decisions about these approximations and developed their own climate models, each of them unique. This development of many independent models makes our confidence in modeling stronger, because despite their differences, all of these models now yield very similar results. We'd only expect this to be the case if the models are successfully taking into account the most important climate factors.

Today's climate models are fantastically detailed, and they reproduce the actual climate of the past century with remarkable accuracy. Indeed, the modern models work so well that scientists can use them to conduct "experiments" in which they ask what would happen if this or that were different than it is. Figure 2.7 shows an example of the power this approach provides. The brown curve shows temperatures over the past century and a half as predicted by the best available climate models, which take into account both natural factors affecting climate, such as volcanic eruptions and changes in the amount of sunlight reaching Earth, and human factors, such as the increase in the carbon dioxide concentration from the burning of fossil fuels. Notice that these models provide an excellent match to the general trends in the real data (black curve). In contrast, when the models are run without the human factors, the results (the green curve) do not agree with the observed warming of the past few decades. The fact that we get a close match between the models and reality only when changes in both natural *and* human factors are included gives us great confidence that human factors are the cause of the recent warming.

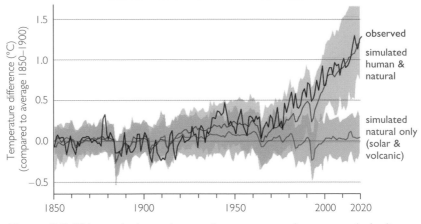

Figure 2.7. This graph shows the excellent agreement between today's climate models (brown curve) and actual temperature changes (black curve), and the lack of agreement when the models consider only natural factors (green curve). Conclusion: Today's climate models work extremely well and demonstrate that global warming is caused by human factors such as the rising carbon dioxide concentration.

Note 1: The model curves are averages of many independent models of global warming; the wider swaths represent the range of predictions found by the different models. Note 2: Bloomberg Business created an outstanding animation set to illustrate the ideas in this graph, which you can see at bloomberg.com/graphics/2015-whats-warming-the-world/. Source: Intergovernmental Panel on Climate Change (IPCC) 2021, Summary for Policy Makers, Figure SPM.1b.

Q What's the bottom line for Skeptic Claim 2 ("the warming is natural")?

There are no known natural factors that could account for the substantial global warming of the past century. Observations definitively rule out the Sun as the cause, and we've seen that today's sophisticated models match up extremely well with observations of the actual climate — but only when we include the human contributions to global warming, not natural factors alone. This fact gives us great confidence not only that the models are on the right track but also that human activity is the cause of most or all recent global warming.

Skeptic Claim 3: It's Warming Up, Humans Are Causing It, but It's Nothing to Worry About

Our third general type of skeptic claim accepts the reality of global warming and the fact that it is human-caused, but argues that this is not a great cause for concern. This claim tends to come in three major subtypes. Some skeptics use the fact that the climate has varied naturally in the past to suggest that we needn't worry about climate change today. Other skeptics claim that future warming will be less significant than most models predict, in which case there is less cause for concern. A few skeptics suggest that warming may even be beneficial to us. Each of these variations of Claim 3 is worth looking at individually, especially since that will give us the opportunity to consider important climate issues that we have not yet had a chance to discuss.

Skeptic Claim 3, Version 1: Natural Climate Variability

There is no question that Earth's climate varies naturally over time, and skeptics have seized on this fact in two major ways. Some have used it to argue that the current warming might simply be part of a natural cycle, but we've already discussed the fact that natural factors are unable to explain this warming. A somewhat more legitimate debate concerns the question of how much danger we face from the current warming, given what we know about past climate change. So let's investigate.

Earth has had many ice ages that had nothing to do with humans; how do those natural climate changes compare to what we're experiencing today?

Geological data do indeed tell us that Earth has cycled in and out of ice ages in the past, and humans obviously did not cause those changes. We can study changes in Earth's average temperature over the past 800,000 years with the very same ice cores (see figure 1.9) used to measure past carbon dioxide concentrations.[6] Figure 2.8 shows the temperature record (red) along with the carbon dioxide record (blue) that you saw previously in figure 1.10. Notice that temperatures have fluctuated significantly. The cool periods are ice ages, and warm ("interglacial") periods come in between them. Notice that, just as we would expect, the warm periods match up with higher carbon dioxide concentrations and the cool periods with lower carbon dioxide concentrations, providing further confirmation that our basic 1-2-3 science is correct.

The skeptics (in our claim 3, version 1 category) point to these large natural changes to suggest that the changes we are causing today are nothing to worry about. But consider these key points:

---

6 The temperature information is derived from careful measurements of isotope ratios (particularly of oxygen-18 to oxygen-16 and deuterium to hydrogen) in trapped air bubbles, cross-checked against other available data. A good starting point for further detail is this NASA webpage: earthobservatory.nasa.gov/Features/Paleoclimatology_IceCores/.

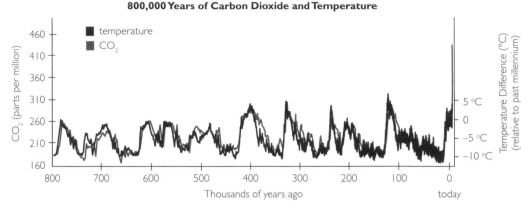

Figure 2.8. This graph shows changes in the temperature and the carbon dioxide concentration over the past 800,000 years. Notice the close correlation, just as we would expect from our basic 1-2-3 science.

Source: Data from the European Project for Ice Coring in Antarctica (EPICA).

- The current temperature is already near the highest it has been in the past 800,000 years. Given that the current carbon dioxide concentration is now some 50% higher than at any other time in that period — and is rising rapidly — it would seem that we should be very concerned about how much higher the temperature will rise.
- Although the figure makes it look as if the natural onset of warm and cool periods occurred fairly rapidly, when you consider that the graph shows 800,000 years, you'll realize that "fairly rapidly" still means "over centuries." In contrast, the changes we are causing today are happening over decades. Again returning to the chapter 1 opening quote from Margaret Thatcher, what we are doing now is "new in the experience of the Earth."

To summarize, while it's true that the climate changes naturally, we are causing changes of an unnatural degree at an unnatural rate. It's hard to see how anyone could take any comfort from these facts.

Q Why does the recent carbon dioxide rise look so much larger than the temperature rise in figure 2.8?

It's primarily an artifact of the scaling used on the graph. The 800,000-year time scale makes the temperature ups and downs of the past 10,000 or so years fall so close together that it is difficult to see the significant rise (more than 1°C) of the past century; the vertical scaling needed to line up the carbon dioxide and temperature data also contributes to making the carbon dioxide rise look more substantial. If you look back at figure 2.4, you can see that the temperature rise actually tracks the carbon dioxide rise quite well.

However, there is also one real issue involved, which brings up an important point about how much warming is already "baked in" to the climate system. Because most of the excess energy currently being trapped by the strengthening greenhouse effect goes into warming the oceans, and water takes much longer to respond to changes than air, we expect there to be a lag time between the current increase in the carbon dioxide concentration and the increase in the global average surface temperature. There is some uncertainty about the lag time, but recent models suggest that the warming tends to lag the increase in carbon dioxide by about a decade.[7] This implies that even if we could cut our greenhouse emissions to zero immediately, the world would continue to get warmer for roughly another decade.

Q What causes the natural cycles of ice ages and warm periods?

The observed pattern of ice ages and warm periods lines up very well with a pattern of small, cyclical changes — called *Milankovitch cycles* (after the scientist who first recognized them) — in Earth's axis tilt and orbit that arise from gravitational effects of the Sun, Moon, and other planets. But there's a very important point that goes along with this: By themselves, the changes that would occur as a result of the Milankovitch cycles are *not enough* to explain the large temperature swings that have occurred. Instead, these cycles are "triggers" that initiate feedback processes that amplify the temperature changes.

Here's how the process is thought to work when a warm period begins: The changes due to the Milankovitch cycles slightly increase the amount of sunlight warming Earth and the oceans. This warming causes the oceans to release some of their dissolved carbon dioxide into the atmosphere.[8] The extra carbon dioxide in the atmosphere causes additional warming, which in turn leads to more evaporation from the oceans. The added water vapor further amplifies the warming, because water vapor is also a greenhouse gas. In addition, the warming leads to melting that reduces the amount of ice around the poles (this is the glacial retreat at the end of an ice age), which lowers Earth's reflectivity and therefore causes our planet to absorb more sunlight and warm even more. To summarize, a small initial warming (triggered through the Milankovitch cycles) leads to a chain of reinforcing feedbacks that leads to a much larger warming.

An opposite set of changes amplifies the cooling side of the Milankovitch cycles. When a cycle initiates a slight cooling, the cooling causes the oceans to absorb carbon dioxide from the atmosphere. This weakens the greenhouse effect, further cooling our planet, which leads to even less evaporation from the oceans and an expansion of the reflective polar ice caps. Together, the added ice and reduced water vapor in the atmosphere amplify the cooling until Earth plunges into an ice age.

---

7 See, for example, Zickfeld and Herrington, 2015, at iopscience.iop.org/article/10.1088/1748-9326/10/3/031001/pdf, and Dvorak et al., 2022, at doi.org/10.1038/s41558-022-01372-y.

8 This release occurs because warmer water generally holds less dissolved gas, an effect you can confirm by popping open a warm can of soda, which will release gas more quickly than a cold can. Additional note: This fact may make you wonder why the oceans are currently absorbing carbon dioxide as our planet warms, and the answer is that it's a short-term gain arising from the great rate at which human activity is releasing carbon dioxide into the atmosphere. Ultimately, we expect the oceans to release more carbon dioxide as Earth warms, which would tend to exacerbate the problem of global warming over the long term.

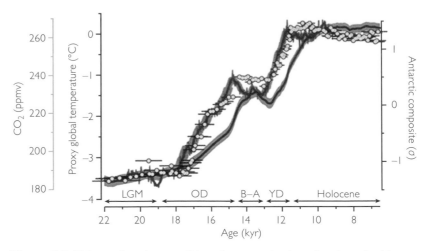

Figure 2.9. This graph makes possible a close examination of carbon dioxide concentration (yellow dots), Antarctic temperature (red curve), and global average temperature (blue curve) during the warming period that began about 20,000 years ago ("kyr" means "thousands of years"). Notice that while the Antarctic temperature changes come slightly ahead of the carbon dioxide changes, the global temperature changes do not.

Source: J. D. Shakun et al., *Nature* 484 (Apr. 5, 2012): 49–54.

Q I heard that the temperature changes in the ice core record precede the carbon dioxide changes. Doesn't this suggest that you have cause and effect backward?

It is true that, in some cases, temperature changes in the ice core record appear to have preceded a rise in carbon dioxide, but this does not in any way change our understanding of the cause and effect. In fact, it's completely consistent with the idea that, as discussed above, the Milankovitch cycles trigger small changes in climate that are then amplified by feedbacks. More specifically, the feedback processes mean that once the Milankovitch cycles initiate a temperature change, both the temperature and the carbon dioxide concentration will rise or fall together, and at any given moment or any given place on Earth, one or the other may change first.

   Further evidence that the cause and effect are well understood comes from a closer look at data from the end of the last ice age. The details are beyond the scope of this book, but the brief summary is as follows. The ice cores that show a slight lead in temperature changes compared to the carbon dioxide changes come from Antarctica, which means they reflect the temperature changes that occurred over the Antarctic ice sheet. However, scientists have other ways to study past temperatures, such as by drilling into sediments in lakes or the ocean floor, and these make it possible to measure past temperature changes in many places around the world. This work is fairly difficult compared to ice core measurement, but figure 2.9 shows what scientists found for the end of the last ice age. The yellow dots mark the carbon dioxide concentration, the red curve indicates Antarctic temperatures (as inferred from ice cores), and the blue curve represents global average temperatures (as inferred from measurements made in many places around the world). Notice that while the Antarctic temperature rise came very slightly ahead of the carbon dioxide rise (which, as stated above, is unsurprising), the global temperature rise came *after* the carbon dioxide rise — completely undercutting any claim that cause and effect are backward.[9]

What about the Medieval Warm Period, when Greenland was "green"? Doesn't that mean that we've been through much warmer periods in the recent past?

Let's start with the first of the two questions here: It's true that the Vikings built settlements on the coast of Greenland beginning about a thousand years ago, during what is known as the Medieval Warm Period (roughly 950 to 1250 CE), when reduced Arctic sea ice made journeys to Greenland much easier than they were in the centuries before and after. But even at that time, Greenland was hardly "green"[10] — the vast bulk of Greenland has been covered by an ice sheet for at least several hundred thousand years. The Viking colonization never occupied more than a few coastal regions.

Now we turn to the second and more important question, which is whether the Medieval Warm Period is relevant to current global warming. The answer is a strong and definitive "no." The reason is simple: Even though there *was* a Medieval Warm Period that aided the Viking settlements, the amount of warming at the time pales in comparison to the warming going on today. Figure 2.10 shows the data from numerous independent scientific studies (each in a different color), along with recent temperature data (red). Notice that while the different studies do not all agree perfectly for times further in the past, they do all agree that today's temperatures are significantly higher than those of the Medieval Warm Period.

In fact, the evidence is even stronger than that shown in figure 2.10, because figure 2.10 includes only Northern Hemisphere temperatures. This is important because careful studies indicate that the Medieval Warm Period was primarily a regional phenomenon. Global data indicate little if any overall warming during this period (figure 2.11) and confirm that the recent warming far exceeds anything that happened in the past 2,000 years; longer-term records, like those in figure 2.8, show that we are now living in Earth's warmest period of at least the past 100,000 years.

---

9   The delay between the carbon dioxide rise and the global temperature rise is thought to be due to complexities of ocean circulation; for details, see skepticalscience.com/skakun-co2-temp-lag.html.
10   According to histories written not long after the colonization, the name "Greenland" was primarily a marketing ploy by the famed Viking Erik the Red, who believed it would encourage other people to make the journey there.

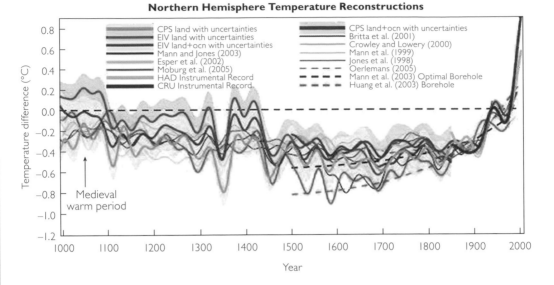

Figure 2.10. This graph shows more than a dozen different independent data sets, all pointing to the same basic fact: Recent temperatures (solid red) are significantly higher than temperatures during the Medieval Warm Period or any other time in the past 1,000 years. This graph is nicknamed the "hockey stick" because it looks kind of like a hockey stick lying on the ground with its tip pointing up at the right.

Source: M. E. Mann et al., *PNAS* 105, no. 36 (2008): 13252–13257.

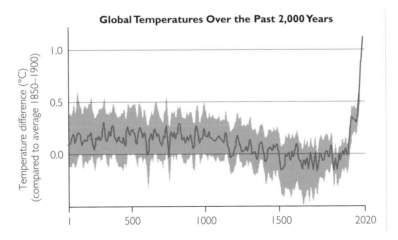

Figure 2.11. This graph shows changes in the *global* average temperature (as opposed to the Northern Hemisphere temperatures plotted in figure 2.10) over the past 2,000 years; the solid line represents the best estimate, and the swath indicates the range of uncertainty. The Medieval Warm Period no longer shows clearly, telling us that it was primarily a regional rather than a global phenomenon.

Note: Data prior to 1880 are based on ice cores and other methods for inferring past temperatures; recent data are based on temperature records as shown earlier (e.g., figure 1.15). Source: IPCC 2021, Summary for Policy Makers, Figure SPM.1a.

Q Wait — didn't I hear that the hockey stick graph (figure 2.10) has been discredited?

You may well have heard this, since it is frequently repeated in places like the *Wall Street Journal*'s op-ed pages, *but it is not true.* The original version of the "hockey stick" was published by climate scientist Michael Mann in 1998, and he used only a single data set. Skeptics jumped on it, claiming all kinds of reasons why the data should be doubted. Scientists took the skeptic concerns seriously and did what scientists do: They investigated in much more detail. Indeed, the reason you see so many data sets — which represent independent sources including tree rings, corals, stalagmites, ice cores, and more — in figure 2.10 is that the scientific community went to great lengths to try to either confirm or refute Mann's original "hockey stick." Keep in mind that every curve you see in figure 2.10 represents many years of fieldwork and careful research by a substantial group of scientists, who often put their lives on the line to collect the data in remote and dangerous locations. As you can see, these additional studies clearly confirm Mann's original conclusions. Still not mollified, the skeptics convinced Congress to ask the National Research Council (NRC) to investigate those conclusions. The NRC report, published in 2006, concluded that the graph and the data were fully valid. Additional research since that time has only further strengthened the case for the validity of the "hockey stick."[11]

Q **You've looked back 800,000 years, but I've heard that if you go back further, both carbon dioxide levels and temperatures were significantly higher than they are today. What do you say to that?**

Although it's more difficult to figure out what the climate was like at times further in the past, evidence does indeed indicate that there have been times when Earth was much warmer (and times when it was much colder) than anything we see in the 800,000-year record. For example, during much of the dinosaur period, evidence indicates that the global average temperature was significantly warmer than it is today, and the carbon dioxide concentration may have been well above 1,000 parts per million, more than double today's concentration. But I do not find these facts the least bit comforting; quite the opposite, as they seem to me to suggest that the current warming could be even more devastating than most scientists generally assume.

Let's start by considering the implications of the warm temperatures of dinosaur times. These temperatures were high enough that there were no ice caps in either the Arctic or the Antarctic (figure 2.12), suggesting that a warming that brings back the temperatures of those times would cause the ice caps to melt completely. If that happened

---

11 Two great sources for learning more about this issue are the NRC report (nap.edu/catalog/11676/surface-temperature-reconstructions-for-the-last-2000-years) and Michael Mann's book *The Hockey Stick and the Climate Wars* (Columbia University Press, 2012).

Figure 2.12. This painting shows Antarctica as it may have looked about 70 million years ago, when our planet was so warm that there were no polar caps and the carbon dioxide concentration was probably above 1,000 ppm.

Source: Artwork by James McKay, University of Leeds, from V. Bowman et al., *Palaeogeogr., Palaeoclimatol., Palaeoecol.* 408 (Aug. 15, 2014): 26–47.

(a possibility we'll discuss in more detail later), sea level would rise so much that every coastal city in the world — not to mention most of Florida, Texas, and other low-lying coastal regions — would be deep under water. (Indeed, sea level during portions of the dinosaur period was more than 200 feet higher than it is today.)

As to the carbon dioxide concentration of the distant past, I'll make two points. First, while a carbon dioxide concentration of 1,000 parts per million sounds very high compared to today's 420 (as of 2023), extrapolating from figure 1.10 shows that at the current rate of increase, we would surpass that level by around the year 2200, which is fairly soon on the time scale of human history. Second and much more important, you may hear skeptics claim that the fact that life thrived when the carbon dioxide concentration was much higher than it is today is proof that life can thrive under such conditions. Well, it is — but it's only proof for the species that were living at the time and therefore were adapted to those conditions. There is no reason to think that today's plants and animals would thrive similarly, because today's life is adapted to today's much lower carbon dioxide levels. Moreover, the species that thrived in those past times had millions of years to adapt to those conditions. Many or most of today's species will have difficulty adapting to the climate changes expected to accompany the rapid rate at which we are now raising the carbon dioxide concentration. Overall, it seems incredibly naïve to imagine that the thriving life of past warm periods should give us any comfort at all about the dangers of global warming.

Q Has the climate ever changed as fast as it is changing today?

Yes, but again, this doesn't provide much cause for comfort. The most sudden known climate change in Earth's history was initiated by the impact of an asteroid (or comet) about 66 million years ago, known as the K-Pg (short for Cretaceous–Paleogene) event, and this climate change led to the extinction of the dinosaurs and of most other species living on Earth at the time. Slightly less sudden climate change (also due to natural events) has been implicated in other mass extinctions, including the "end-Permian" mass extinction of about 252 million years ago, in which more than 90% of all plant and animal species went extinct.

The closest known analogue to today's climate change is probably what is called the Paleocene-Eocene thermal maximum (PETM), which occurred about 55 million years ago. The PETM saw a huge increase in the carbon dioxide concentration, possibly to as high as 2,000 parts per million,[12] leading to a global temperature rise of 5–8°C (9–14°F). However, this change took place over a time period estimated to have been between about 20,000 and 50,000 years, which means it was much more gradual than today's global warming. Even so, it apparently led to many extinctions, changes in sea level, changes in ocean currents, and other effects that we would consider to be highly detrimental. In that sense, the fact that we are now changing the planet at a faster rate than during the PETM is just one more reason why we should be very concerned about the implications of the current warming.

Skeptic Claim 3, Version 2: The Reliability of Models

We next turn to the skeptics who argue that fears of global warming are overblown by disputing the reliability of climate models. Could these skeptics be correct? The first thing that any scientist will tell you about modeling is that it's not easy. As an old saying goes, "Prediction is hard, especially about the future." But hard is not the same as impossible, and as we've already discussed, today's sophisticated climate models do an excellent job of "predicting" the past and present climate (see figure 2.7). With that in mind, let's look in a little more detail at the common skeptic claims about modeling and why the vast majority of scientists find these claims unconvincing.

Q Today's models can't even predict the weather more than a few days out, so how could they possibly predict the effects of global warming many years from now?

This common question has a very simple answer which relates to the fact that *weather* and *climate* are not the same thing:

---

12 The precise cause of this huge carbon dioxide release is not yet known, though it was probably tied to a period of unusually strong volcanic activity. A good starting point to learn more is the BBC podcast at bbc.co.uk/programmes/b08hpmmf.

- *Weather* refers to the ever-varying combination of winds, clouds, temperature, pressure, and humidity that makes some days hotter or cooler, clearer or cloudier, or calmer or stormier than others.
- *Climate* refers to the *average* of weather over many years. For example, we say that a desert has a dry climate, even though it may have some days with a lot of rain or snow.

It is always easier to make predictions about averages than about individual outcomes. For example, we cannot accurately predict whether a single coin toss will land heads or tails, but we *can* predict that if we toss many coins, close to 50% of them will land heads. The situation with weather and climate is exactly the same: It's much more difficult to predict the short-term variability of weather than the long-term average that represents climate.

In other words, the fact that it's hard to predict the weather is *completely meaningless* in considering our ability to predict the climate. The evidence that we can predict the climate comes from the fact that today's sophisticated climate models do an excellent job of matching up with the present climate reality, giving us confidence that they should be similarly good at predicting the future climate.

It's also worth noting that when I say that today's models do an excellent job of matching reality, I mean much more than just the global average temperature. Today's models make regional predictions, and these regional predictions also match up to reality quite well. For example, as we'll discuss more in the next chapter, climate models have predicted numerous regional changes that appear to be occurring as expected, such as increases in droughts in the southwestern United States, increased flooding in regions of Asia, and much more. To ignore the insight provided to us by climate models simply does not make any sense.

OK, but couldn't the models still be overestimating the future warming?

This question brings us to an issue called the *climate sensitivity*, which describes how much we can expect the climate to change (how "sensitive" it is) in response to future changes in the carbon dioxide concentration. In essence, some skeptics argue that today's climate models are overestimating the climate sensitivity, in which case they would also be overestimating the future warming.

In principle, it's always possible that the models could be missing some mitigating factors that would lower the climate sensitivity. However, any such factors would have to have some rather strange properties. In particular, because these factors would by definition be factors that are not considered by current models, they would have to be both *insensitive* enough that they haven't caused major failures in the model results through the present time and *sensitive* enough to make a major impact on the model results for the future. We cannot completely rule out the possibility that such factors exist, but these odd properties make it seem rather unlikely.

Q **What about the difficulty of modeling the effects of clouds?**

Clouds are very complex and their effects are still not fully understood, and many skeptics have used this fact to make an argument that begins with something that is almost certainly true: Because global warming makes the world warmer, it leads to increased evaporation and more clouds. However, the skeptics then go on to assume that because clouds reflect sunlight, the additional cloud cover will serve as a self-regulating (negative) feedback that will slow or stop any additional warming. But this latter claim ignores two related effects. First, the additional evaporation that leads to the additional clouds also means there is more water vapor in the atmosphere, and the warming effects of this water vapor (a greenhouse gas) will tend to counter the cooling effects of the clouds. While there is some legitimate debate over which effect is stronger on short time scales (years to a few decades), our understanding of ice age cycles leaves little doubt that water vapor's amplifying role will dominate over longer time scales. Second, clouds are not the only thing that matters to the total reflectivity of Earth. For example, as we'll discuss in the next chapter, the loss of Arctic sea ice caused by global warming reduces reflectivity (because liquid water reflects less sunlight than ice). Most scientists suspect that the combination of the warming effects of added water vapor and reduced ice cover will overwhelm the cooling effects of additional clouds.

Q **Is there any other evidence through which we can gain confidence in the predictions of climate models?**

Yes, and it comes from looking at the predictions of older climate models. To understand this idea, remember that while scientists test new models by looking at how well they match up with the past and present climate, in future years we will be able to see how well the predictions of these models stack up against reality. Extending this idea back in time, we can now check the predictions of models that were run in decades past.

The results are very encouraging for the reliability of models. Figure 2.13 shows one such set of results. As you can see, the predictions of these two-decade-old models have turned out to be quite accurate, and similar studies show that still older models have also been quite successful.[13] Given that today's more sophisticated models incorporate insights based on what the past models got right and what they

---

13 For details on how figure 2.13 was made, see climate.nasa.gov/news/2943. For discussion of even older models, see carbonbrief.org/analysis-how-well-have-climate-models-projected-global-warming.

2

The Skeptic Debate

Forecast Evaluation for Models in 2004

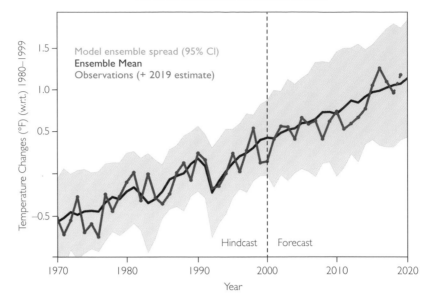

Figure 2.13. This figure compares the predictions of models run in 2004 with actual temperature data (red), both for times for which data were available ("hindcast") and for times for which the models made predictions ("forecast"). The black curve ("ensemble mean") is the average prediction of more than a dozen different models; the shaded swath shows the range (with 95% confidence) of predictions by the different models. Notice that the forecasts have proven to be quite accurate.

Note: This graph uses a different zero line than our previous model graphs and is labeled with temperature changes in °F rather than °C. Source: Gavin Schmidt, NASA.

got wrong, we can expect today's predictions to be even more reliable going forward. In fact, the simplicity of the underlying science means that even a rudimentary model can capture the key ideas. Recall from chapter 1 (see page 12) that the Swedish chemist Svante Arrhenius created a simple climate model[14] way back in 1896. He used this model to predict the warming effect of a doubling of the carbon dioxide concentration, and his result agreed remarkably well with the results of today's best climate models.

To summarize, we have plenty of evidence supporting the general reliability of models, and little reason to think that they are missing any important mitigating factor. So while we can never be 100% certain that the model predictions are accurate, we ignore them at our peril.

---

14   Although Arrhenius's model was "simple" in the sense that the calculations could be done with pencil and paper, getting his result required more than 10,000 calculations that he did over a period of close to a year.

What if I still don't trust the models?

OK, let's say you want to ignore all the evidence for the reliability of models and stick to the skeptic claim that, because models can never be perfect, we shouldn't trust them at all. Well, that's not very scientific of you, but let's go with it for the moment.

Whether or not you believe the models, you still have to make decisions about what, if anything, we should do about global warming. If you don't want to consider the models, the only viable alternative is to base your decisions on the actual data. So let's see what the data tell us. Look back at figure 2.8, where you can see how temperature and carbon dioxide concentration have changed together in the past. Notice that (1) over the past 800,000 years, the largest upward swings in carbon dioxide concentration have been from about 180 to 290 parts per million, which is an increase of about 60%; and (2) these 60% increases in carbon dioxide concentration have been accompanied by temperature increases larger than 10°C (18°F).

Putting those two facts together, the past data seem to suggest that a 60% increase in the carbon dioxide concentration would cause warming of 10°C or more. This is substantially *greater* warming than any of the models suggest for the rest of this century, even though current trends imply that we would reach double (a 100% increase) the preindustrial carbon dioxide concentration before the century ends. In other words, the warming you would project from actual past data is worse than what the models are predicting.

There's one last way in which some skeptics try to argue this point, which goes to the fact (shown in figure 2.8) that the world was already near its warmest of the past 800,000 years before humans started to affect the climate.[15] Perhaps, say the skeptics, this might mean that the world just can't get much warmer. There is probably a kernel of truth underlying this idea, since the fact that the models predict less warming than we would project from the data alone suggests that an already-warm climate is less sensitive to further increases in the carbon dioxide concentration. However, it cannot be the whole story because, as we've discussed, Earth has been much warmer at some times in the past (for example, in dinosaur times and during the Paleocene-Eocene thermal maximum). Clearly, there is plenty of room for the warming to continue if we keep adding more greenhouse gases to our atmosphere.

---

15 More specifically, the world has been in a warm interglacial period for about the past 12,000 years (what geologists call the Holocene epoch), and most of the human-caused warming has occurred only within the past century or so.

Skeptic Claim 3, Version 3: Benefits May Outweigh Risks

We turn next to a version of our skeptic claim 3 that goes beyond disputing the magnitude of the threat by arguing that the effects of added carbon dioxide and global warming are actually *good for us*. This is a rather remarkable assertion in light of all the detrimental effects that we are already seeing (and will discuss in the next chapter), but it has gained some prominence in recent years. Here, we'll take a brief look at the two major benefits claimed by these folks, which revolve around (1) Arctic warming and (2) effects on agriculture.

Q Is the melting of Arctic ice good for us because it opens up the Arctic sea?

The melting of Arctic sea ice is already causing a rush for Arctic riches (such as minerals, oil, and other resources that lie under the Arctic Ocean) and for faster shipping routes among northern countries. By themselves, these things might seem to be beneficial to the global economy. But the key words are "by themselves," because they don't occur in isolation. Instead, they are consequences of having less sea ice, and that appears to be a very detrimental development overall; read ahead to chapter 3 (page 87) to see why. To summarize, any economic benefits from less ice in the Arctic Ocean are almost certainly outweighed by the negative global effects that go along with them.

Q Will more carbon dioxide be beneficial to plant growth and agriculture?

It's true that carbon dioxide is "plant food" that is necessary for plant survival and growth. But just as you won't necessarily get healthier by increasing the amount of food you eat, the fact that plants need carbon dioxide doesn't automatically mean that more is better. So to support their claim of agricultural benefits from more carbon dioxide, the skeptics generally cite three types of "evidence." First, they point to the thriving plants under the high carbon dioxide concentrations of dinosaur times. But we've already seen how this argument falls apart when you consider the time needed for adaptation (see the discussion that begins on page 53).

Second, they cite small-scale experiments from a couple of decades ago that showed some benefits from higher carbon dioxide concentrations for crops such as soybeans and rice. However, more recent and larger-scale versions of these experiments have demonstrated that any such benefits are small to nonexistent.[16] Moreover, even if individual crops might benefit in isolation, this benefit is likely to

---

16 These types of experiments go by the general name "Free-Air Carbon dioxide Enrichment" (FACE) experiments. You can learn about them by searching that term, but a good starting point is this U.S. Department of Energy website: ess.science.energy.gov/face.

be overwhelmed by broader ecological effects. For example, agriculture might still suffer due to the expected increase in crop-damaging weather events. More generally, remember that plants and animals are adapted to *local* climates. If the climate changes slowly, species can adapt or migrate to survive. But if the climate changes faster than they can adapt or migrate, they will die out or be outcompeted by other species. The rapid rate of climate change today makes it likely that there will be great changes to the entire ecosystem. It seems a great stretch to imagine that such changes might somehow make the ecosystem better for agriculture than it was during the centuries and millennia in which our civilization arose.

Third, they cite a "greening of the Earth," which refers to research indicating that there has been a significant increase in total plant mass over the past few decades. Once again, this might seem beneficial in isolation, but not when considered in the context of the many other climatological and ecological changes that are happening. In this case, the skeptics simply choose to ignore the context. For example, they cite real studies to support their benefit claims, but omit the fact that these studies actually reach the conclusion that any benefits are outweighed by other negative impacts.[17] They also completely ignore studies showing that the greening has been slowing and is unlikely to continue (see, for example, bg.copernicus.org/articles/18/4985/2021/), as well as the fact that any rapid change in the ecosystem is far more likely to be damaging than beneficial. To sum up, any claims that the "greening" of the Earth is automatically good for us are based on a highly selective and misleading review of the actual research.

Q What's the bottom line for Skeptic Claim 3 ("nothing to worry about")?

We've looked at three general ways in which some skeptics try to argue that global warming is nothing to worry about while accepting that it is happening and is human-caused, and in each case I've pointed out serious flaws in the arguments. But let's take a moment to consider what it would mean if the arguments actually had any merit. Even in that case, these skeptic arguments would at best offer a *possibility* that global warming might not be as bad as climate models predict. In other words, these skeptics are essentially advocating that we continue doing an "experiment" on our planet in which we keep adding greenhouse gases to the atmosphere without being sure of the consequences. It may be true that we can't completely rule out the chance that this experiment would result in more pros than cons, but given what we

---

17   For details on this type of misinterpretation by the skeptics, see my article at jeffreybennett.com/debate. I also recommend this article on the origin of the "benefits" claims: yaleclimateconnections.org/2020/09/video-origin-of-the-myth-that-global-warming-good-for-agriculture/.

know about the basic science, such an outcome seems highly unlikely. We can only conclude that there is great risk involved in continuing this experiment, which brings us back to the question I asked you to keep in mind at the beginning of this chapter: How much risk are you willing to accept?

Skeptic Claim 4: It's Warming Up, Humans Are Causing It, It's Harmful, but We Can't Do Anything About It Because . . .

We have now reached the fourth major skeptic claim, which accepts both the reality and the danger of global warming, but still argues against taking action.

Q How can skeptics agree that global warming poses danger but still argue against action, and what do you say in response?

The skeptics in this case tend to make one or more of three general arguments. The first argues against moving away from fossil fuels by citing the fact that low-cost energy from these fuels has played a major role in building our modern economy and raising millions of people out of poverty. I agree that we shouldn't give up the benefits we've received from fossil fuels, but there's an obvious flaw in this skeptic argument: It assumes that fossil fuels are the only cost-effective means for obtaining these benefits. As we'll discuss in chapter 4, I don't believe that this is the case.

The second general argument suggests that the costs of dealing with global warming are high enough that more good could be done by applying our efforts in other areas, such as combatting childhood mortality or global poverty. This might sound noble, but I believe it poses a false choice. Instead, as I'll argue in the final two chapters, I believe that the same steps needed to tackle global warming can also offer solutions to global poverty and many other problems.

The third general argument essentially says that it's already too late to stop global warming, so we might as well just accept it and do the best we can to come up with strategies for adaptation as the world changes. This is the very argument of despair that, as I mentioned in the introduction (pages 2–3), is the main reason I decided to write this second edition of this book. The threat to our future posed by global warming is indeed serious, but it is *not yet too late* to address it.

Q What's the bottom line for Skeptic Claim 4 ("we can't do anything about it")?

Given that it is so clearly defeatist to believe that we can't do anything about global warming, you might wonder why anyone would promote this position. Yet many skeptics do, and interestingly, many of them are the very same people who've promoted other skeptic claims in the past. In other words, the "skeptic community" has gradually moved through what my good friend and textbook co-author Nick Schneider has called "the four levels of climate denial":

- Level 1 (denying the data): "The Earth is not warming up."
- Level 2 (denying the cause): "It's warming up, but it's natural."
- Level 3 (denying the consequences): "It's warming up and humans are causing it, but it's harmless or even beneficial."
- Level 4 (denying responsibility): "It's warming up, humans are causing it, and it's harmful, but it's beyond our capacity to do anything about it."

Now that you've read about the skeptic claims, I hope you'll agree that it's time to get out of denial. The sooner we act, the better our chances of limiting the damages in the short term and of ultimately finding a pathway to a post–global warming future.

The Scientific Consensus

I'll now turn to what is often called the "scientific consensus" on global warming, which holds that global warming is real, is human-caused, and poses a major threat to our civilization — and therefore that we should take strong action to limit the future warming. The idea of a scientific consensus is common in virtually all fields of science, but the skeptics have attacked this idea when applied to global warming. I'll therefore close this chapter by briefly addressing the relevant issues.

Q Science isn't a democracy, so why should we pay attention to a consensus?

It's true that scientific facts can't be established by a vote. But there is no single authority that can represent science as a whole, and this means that we have to weigh evidence regarding scientific claims in much the same way as a jury weighs evidence in a criminal trial. Generally speaking, we accept a scientific fact or theory when the evidence in its favor becomes overwhelming. Much as in a criminal trial, it's never possible to reach 100% certainty, so we instead strive to reach a point where we can say that the evidence favors some fact or theory "beyond a reasonable doubt."

In the case of climate change, I'll point you to three major reasons why I believe that the scientific consensus has indeed been established beyond a reasonable doubt:

1. As I hope I've shown you in these first two chapters, the basic 1-2-3 science of global warming is easy to understand and supported by overwhelming evidence, while the skeptic claims fall apart on close examination. You can therefore see the evidence for yourself, without even needing to rely on the voices of the consensus view.

2. Decades ago, when there were many more unanswered questions about global warming, scientists and policy makers recognized a need for an organization that would evaluate the science in great detail. In 1988, the United Nations and the World Meteorological Organization established the International Panel on Climate Change, better known as the IPCC, to conduct comprehensive assessments of the best available science relating to climate change, its impacts and risks, and options for adaptation and mitigation. The IPCC produces detailed reports that represent the combined work of thousands of people, including hundreds of active researchers in climate science, all working in an open and transparent way.[18] These reports in essence represent the scientific consensus at the time of their publication, and the most recent reports have been unequivocal in citing the reality and dangers of global warming.

3. Just in case the above two points are not enough, surveys have shown that more than 97% of scientists who have spent their lives studying the climate accept the consensus view, and the peer-reviewed scientific literature is essentially unanimous in supporting the consensus.[19] Imagine that you visited 100 doctors for advice on some medical question, and 97 of them gave you the same answer while 3 claimed that the other 97 were all wrong. Which set of doctors would you listen to?

A final note on the consensus: You may occasionally hear what I call the "Galileo argument" —namely, that the fact that Galileo was persecuted for his claim that Earth goes around the Sun proves that a con-

---

18   You can learn almost anything you'd want to know about the IPCC and its work simply by visiting the IPCC website (ipcc.ch). I highly recommend that you at least skim through the IPCC summary and "working group" reports available at that website, as doing so will quickly give you a sense of the extreme level of effort and extensive research that has gone into these reports.

19   Multiple surveys have found the approximately 97% value; see skepticalscience.com/global-warming -scientific-consensus-advanced.htm. With respect to the peer-reviewed literature, a 2021 study found that at least 99.9% of published climate science papers were consistent with the consensus view (see Lynas et al., 2021, *Environ. Res. Lett.* 16 114005).

sensus can be wrong. But this is a gross distortion of the reality. Galileo's persecutors were religious and political authorities, not scientists. His scientific work built upon evidence that had already been accumulating and to which he added further evidence that elevated the claim (that Earth is a planet) beyond reasonable doubt. In other words, Galileo actually represented the emerging scientific consensus, while his persecutors represented the skeptics who sought to deny reality.

Q Couldn't the scientific consensus be a result of peer and funding pressure?

The last bastion of the skeptic claims comes down to arguing that the entire climate science community is succumbing to some sort of "group think" bias through which they are missing a truth that only the skeptics can see, presumably because of peer pressure or funding pressure (in terms of getting research grants). But this claim is preposterous. To begin with, remember that our general understanding of the greenhouse effect and global warming goes back to long before today's funding agencies were established — which means the funding followed the research, not the other way around. Moreover, the climate science community includes thousands of scientists from every nation of the world, all of them devoting their careers to hard and detailed work; does it really seem possible that they could all miss the same "hidden truth" (or engage in a conspiracy to hide the truth)? Finally, even if we imagined that the scientific consensus represented by the IPCC and others was somehow a result of peer and funding pressure, that certainly would not be the case for the scientists who worked in the past at companies like Exxon, where any pressure would have been in the opposite direction. Yet as has been recently reported, the Exxon scientists came to exactly the same conclusions that form the consensus view,[20] completely undercutting any claim of group bias or conspiracy.

To summarize, today's understanding of global warming as expressed by the consensus view represents a triumph of detailed scientific examination that has taken place over the course of more than a century and a half through the hard work of thousands of great scientists and careful vetting by many more. No scientific theory can ever be proven beyond all doubt, but the reality of global warming and of the threats that it poses have certainly been established beyond any reasonable doubt.

---

20 You can find the article about the Exxon scientists at science.org/doi/10.1126/science.abk0063.

3 The Expected Consequences

We served Republican presidents, but we have a message that transcends political affiliation: the United States must move now on substantive steps to curb climate change, at home and internationally. There is no longer any credible scientific debate about the basic facts: our world continues to warm, with the last decade the hottest in modern records, and the deep ocean warming faster than the earth's atmosphere. Sea level is rising. Arctic Sea ice is melting years faster than projected.

—William D. Ruckelshaus, Lee M. Thomas, William K. Reilly, and Christine Todd Whitman, former heads of the Environmental Protection Agency under Presidents Nixon, Reagan, George H. W. Bush, and George W. Bush, Aug. 1, 2013 (statement in *The New York Times*)

Global warming has many consequences for the overall climate system, a few of which are noted in the quote above from four past Republican leaders of the Environmental Protection Agency. Indeed, global warming is expected to impact our world in so many ways that it would be impossible to discuss them all in a short book like this. To keep our discussion simple, I'll therefore group the consequences into the following five categories (keep in mind that you may see different groupings in other sources):

- Local and regional climate change
- Storms and extreme weather
- Melting of sea ice
- Sea level rise
- Ocean acidification

In this chapter, we'll briefly look at how and why each set of consequences is occurring. We'll also discuss what current models predict about future warming (because that determines how severe the consequences will become) and the possibility that we might reach some "tipping point" that could amplify some or all of the various consequences.

It's worth noting that this discussion will in some sense change our focus from *global warming* to *climate change*. Although the two terms are often used interchangeably, there is a subtle difference between them. *Global warming* refers to the fact that Earth's global average temperature is rising. *Climate change* refers to the many local and regional changes in Earth's climate that are occurring as a result.

Q Is there a simple explanation for why global warming leads to the five categories of consequences?

Yes, and it is summarized in figure 3.1. Recall that the root cause of global warming is that we are emitting carbon dioxide (and other greenhouse gases) into the atmosphere, but that only about half of this carbon dioxide remains in the atmosphere, while much of the rest is absorbed by the oceans. This fact leads to the first branch point in figure 3.1:

* As shown on the left, the carbon dioxide that remains in the atmosphere strengthens the greenhouse effect.
* As shown on the right, the carbon dioxide absorbed by the oceans dissolves in the ocean water.

Let's now continue down the left branch. Recall that the greenhouse effect acts like a blanket, trapping *energy* in Earth's lower atmosphere and oceans. The stronger greenhouse effect means that more total energy is being trapped. This extra energy not only warms our planet but also has many other effects, including the first four of our five major categories of consequence.

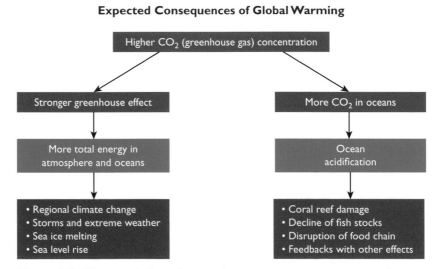

Figure 3.1. This simple flow chart explains why we expect a variety of consequences from what we usually just call *global warming*.

Q Can you give us an analogy to explain why the extra energy trapped by a stronger greenhouse effect will do more than just warm the planet?

Consider heating a pot of water on a stove. The energy from the stove obviously heats the water, but that's not all that it does. Some of the energy makes the gas flame or electricity flow, some of it goes into the surrounding air, and some of it even drives "weather" (more technically, *convection*) in the pot by causing the water to circulate as heated water rises up from the bottom and cooler water sinks down from the top. In the same way, the added energy trapped by a stronger greenhouse effect in the atmosphere not only heats the air but also heats the oceans, melts ice, and drives more weather (and extreme weather).

Our fifth major category of consequence shows up on the right branch of figure 3.1, where the key fact is that the dissolved carbon dioxide undergoes chemical reactions that make the water more acidic. I'll treat this *ocean acidification* as a single category of consequence, but as the figure shows, it has many detrimental effects.

Q **How much total energy is global warming adding to the atmosphere and oceans?**

An astonishingly large amount. Recent measurements indicate that we are causing the total energy of the atmosphere and oceans to increase by approximately 500 trillion joules[1] *each second* — a rate that has roughly doubled just since 2005. To put this in more concrete terms, it is the equivalent of, for example,[2]

- The energy that would be released by detonating eight Hiroshima-size atomic bombs each second.
- The energy that would be released by 1 million lightning strikes each second.
- The energy needed for four places in the world to be struck by major hurricanes at all times.
- The energy needed to power almost 6,000 tornadoes of the most destructive category (F5) each day.

With that much energy being added to the atmosphere and oceans, it should not be surprising that we are seeing noticeable consequences. Indeed, you might wonder why the consequences aren't even greater than they are, and the answer is that most of the energy is going into the gradual warming of the oceans and atmosphere.

---

1 The *joule* is the standard international unit of energy; Americans will be more familiar with energy measured in food calories (1 food calorie = 4,184 joules).

2 A few details about the comparisons: The first two are simply other ways of stating the equivalent amount of energy per second. The third is based on the fact that the energy increase per second is equivalent to the energy released per second by four major hurricanes. The fourth is based on taking the energy released each *day* and calculating how many tornadoes would be represented by the same total energy. FYI, for those who used the first edition of this book: I had based those numbers on the pre-2005 data, which is why they are now doubled.

Q How do scientists measure the energy increase?

There are two major and essentially independent methods. The first takes advantage of the fact that more than 90% of the added heat is going into the oceans, which means that the observed ocean warming (see figure 2.2) can be used to calculate the approximate increase in the total heat content (energy) of the climate system. The second method uses satellite measurements of what is called Earth's *energy imbalance*, meaning the difference between the total energy of sunlight entering Earth's atmosphere and the total energy that is going back out. This difference represents the extra energy that the greenhouse effect is trapping. This second method is somewhat more complex than the first one because the difference varies substantially with time (due primarily to changes in cloud cover, weather, and the El Niño–Southern Oscillation). Nevertheless, careful studies provide a good average value for the energy imbalance.[3] Both methods yield very similar results, giving scientists confidence in their validity.

Local and Regional Climate Change

Recall that as of 2022, the world had already warmed by about 1.1°C (2°F) over the past century. If the only result of this warming were that every day and every place became that much warmer, it probably would be no big deal; it might even sound pleasant if you live in a cold climate today. But the warming is not uniform from day to day or place to place, which brings us to our first category of consequence: local and regional climate change.

Q **How are temperatures changing around the world?**

Figure 3.2 compares recent regional temperatures to their averages from 1951 to 1980. Notice that while almost all regions have become significantly warmer, some regions — particularly in the Arctic — have warmed much more than others. Recall also that nights have generally warmed more than days and winters more than summers.

Q **How are regional temperature changes affecting seasonal cycles?**

The regional warming is having dramatic effects on seasonal cycles, effectively making summers longer and winters shorter.[4] As an example, figure 3.3 shows this change averaged over the contiguous United

---

3 The energy imbalance is usually reported in units of power per square meter, with recent measurements indicating an average imbalance of about 1 watt per square meter (see nasa.gov/feature/langley/joint-nasa-noaa-study-finds-earths-energy-imbalance-has-doubled). If you multiply this value by Earth's surface area, you'll find that it equals about 500 trillion watts, which is the same as 500 trillion joules per second.
4 Keep in mind that global warming does not have any impact on Earth's orbit around the Sun, so when I say that summers are longer and winters are shorter, I'm speaking only of the weather-based definitions of seasons, not of their astronomical definitions based on the solstices and equinoxes.

States. Notice that, compared to the long-term average, the date of the first fall frost (meaning a day with temperatures below freezing) is now coming about seven days later while the last spring frost is coming about six days earlier, for a net of thirteen additional frost-free days.

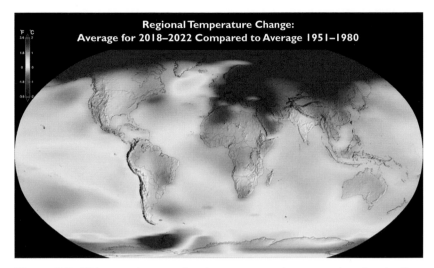

Figure 3.2. This map shows regional temperature changes, comparing the five-year period 2018–2022 to the average from 1951 to 1980. Notice that the warming has been particularly large in the Arctic. Be sure to see the NASA video of these changes (going back to 1880) posted at youtu.be/5LBkKtgS0FI.

Source: NASA Scientific Visualization Studio; you can generate similar maps for any time period at data. giss.nasa.gov/gistemp/maps/.

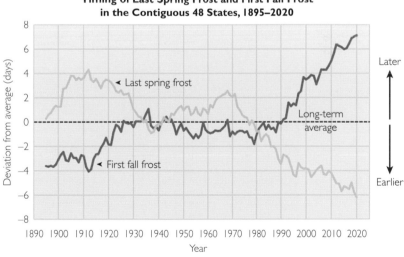

Figure 3.3. This figure shows how the average timing of the first fall frost (orange) and last spring frost (yellow) has changed over the decades.

Source: U.S. Environmental Protection Agency, based on a 2021 update to data from K. E. Kunkel et al., *Geophys. Res. Lett.* 31:L03201.

One fairly obvious consequence of this change is a shorter season for skiing and other winter sports, which can damage communities that depend on these sports. A much more significant consequence is that the longer summers, combined with shorter and warmer winters, tend to promote the spread of insect-borne diseases and crop-destroying pests. This spread can have deadly consequences both for individuals and for the global food supply.

Q How does climate change lead to the spread of pests and disease?

The seasonal changes associated with global warming promote the spread of insects in two major ways. First, warmer temperatures tend to shorten insect growth cycles, allowing insect populations to grow more rapidly. Second, the geographic spread of warmer temperatures means that insects can now survive in regions that used to be too cold for them. As an example, consider mosquito-borne diseases such as Zika, dengue fever, and chikungunya. These diseases propagate when a mosquito bites an infected person and then, after some incubation period, spreads the disease to other people through subsequent bites. Research has shown that warmer temperatures shorten the incubation period, which means the diseases can spread more rapidly, and the geographic changes mean these diseases are now spreading into new regions. Similar ideas hold for other insect-borne diseases and for insects that can destroy agricultural crops.

Q Is the spread of insects tied in with dying forests?

Almost certainly yes (figure 3.4). Vast numbers of trees have been dying in temperate forests around the world, with insects as a major culprit (especially beetles that eat through tree bark). For example, the pine forests of the Rocky Mountains have been suffering from an explosion in the population of pine beetles, which can kill the trees. This increase in the pine beetle population has been tied to the fact that the combined effects of warmer

Figure 3.4. The reddish/gray trees in this photo are dead, killed by the climate-linked spread of pine beetles in the Rocky Mountains. This photo was taken just north of Breckenridge, Colorado.
Source: Wikimedia Commons/Hustvedt.

Change in Annual Precipitation (Rainfall/Snowfall)

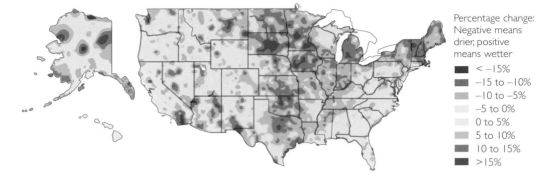

Percentage change: Negative means drier, positive means wetter

■ < −15%
■ −15 to −10%
■ −10 to −5%
□ −5 to 0%
□ 0 to 5%
■ 5 to 10%
■ 10 to 15%
■ >15%

Figure 3.5. This map shows percentage changes in total precipitation over the United States for the period 1986–2015 compared to 1901–1960. Blue/green regions have become wetter, and yellow/brown regions have become drier.

Source: Fourth National Climate Assessment, Volume I, Figure 7.1 (U.S. Global Change Research Program).

temperatures and shorter winters are allowing the beetles to effectively have an entire extra generation of reproduction each year. At the same time, cold snaps that used to kill off beetle larvae are becoming less frequent, further exacerbating the problem. Similar forest die-offs are being observed in many other places around the world.[5]

How and why are precipitation patterns changing?

Climate change is also affecting patterns of precipitation (rain and snow). For example, figure 3.5 shows how the total amount of precipitation over different regions of the United States has changed in recent decades. Notice that while most regions have become wetter, some have become drier, particularly in the southwest. Similar changes are occurring around the world.

You might wonder how global warming can cause both increased wetness and increased dryness (depending on the region), but it's easy to understand if you think about the effects of warmer temperatures. Globally, warmer temperatures mean more evaporation and hence more total moisture and precipitation, which explains why most regions are getting wetter. But the greater evaporation can also cause more and faster drying, which means some areas can become drier. Note that although there are some exceptions, the general trend is that wet regions get wetter and dry regions get drier, making it relatively easy to predict the pattern that your own local region is likely to experience.

5 A good starting article for learning more about the climate-related spread of bark beetles and its consequences is this one from the Yale School of the Environment: e360.yale.edu/features/small-pests-big-problems-the-global-spread-of-bark-beetles.

3 The Expected Consequences

Q Is climate change causing species extinctions?

You are probably aware that human activity has been driving species to extinction at such a high rate that it is sometimes said that we are in the midst of a "sixth mass extinction," so named because the fossil record shows evidence of five major mass extinctions in the past (including the dinosaur and end-Permian extinctions; see page 55). The vast majority of today's extinctions are probably *not* directly caused by climate change, but instead are tied to habitat destruction and overhunting or overfishing. However, climate change is probably playing some role already — and may play a larger one in the future — for a simple reason: Local climate change can significantly disrupt ecosystems, which can drive species to extinction if they are unable to adapt or migrate fast enough to keep up with these changes.

Q Is climate change causing an increase in wildfires?

It's hard to miss the frequent news reports of devastating wildfires (figure 3.6) that have been occurring in many places around the world, often in places where you might not expect them (for example, Alaska and Siberia) and at times of year when wildfires used to be rare. Indeed, not far from where I live in Colorado, more than 1,000 homes were destroyed when the "Marshall fire" rampaged through suburban neighborhoods that most people assumed were safe from fires, and it did this on the last two days of December (2021), when it would have seemed much more likely that the ground would be covered by snow.

Figure 3.6. We expect wildfires to become more common in regions that are becoming drier and hotter with time, and data indicate that this is indeed occurring in many places around the world.
Source: U.S. National Forest Service, photo of the Colorado High Park wildfire.

Beyond the news reports, data support the fact that wildfires are burning much more area than they did in the past.

But is climate change to blame? It would certainly seem to make sense, since most of the increase in wildfires is occurring in regions that are becoming drier (and hotter), and these fires are often exacerbated by factors such as the increased numbers of dead trees, which we've already tied to climate change (see figure 3.4). That said, many other (non-climate) factors are also known to be at play, such as human-caused ignition (sometimes accidental and sometimes deliberate) and the fact that past fire suppression efforts have made many forests unnaturally dense, which means they have much more fuel to burn.

In decades past, these complications meant that while most scientists were likely to say that climate change was *probably* a significant factor in the increased wildfires, they were reluctant to express any higher level of certainty. More recently, however, many scientists have put a great deal of effort into what are called *attribution studies*, in which they seek to tease out the effects of climate change from other effects. Figure 3.7 shows the results of one such study that looked at wildfires in the western United States. As you can see, the study concluded that about half of the observed increase in area burned by wildfires would have occurred even without climate change, but the other half is attributable to climate change.

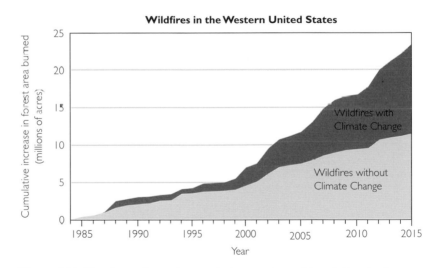

Figure 3.7. This figure shows the cumulative area burned by wildfires in the western United States from 1984 through 2015, compared to what would have been expected due only to natural conditions. The wedges represent what attribution studies have concluded about the portion of this burn that would have occurred even without climate change (lower wedge) and the portion attributable to climate change (upper wedge).

Source: Fourth National Climate Assessment, Volume 2, Figure 25.4, adapted from Abatzoglou and Williams, 2016.

Q **How does attribution science work?**

The basic methodology involves creating two (or more) related models of some event (such as a wildfire or storm), with one model predicting what would have occurred in the absence of climate change and the other including climate change. The differences in the model results provide an estimate of how much climate change affected the event. As with other types of modeling, the accuracy of the modeling used in attribution science depends both on the quality of the input data and on how well we understand the various effects that can contribute to the event being modeled. In general, we expect the models to be better when there is a longer data record. For example, it is easier to do attribution science on floods that occur in places where there is a centuries-long record of past flooding.

Attribution studies are becoming increasingly important in climate science, and improvements in modeling mean that they are also becoming increasingly reliable. That said, because these studies still have a great deal of uncertainty, in this book I'll generally focus more on the actual data, except in cases like that of the wildfires, where there is no other clear way to separate the effects due to climate change from the other human-caused effects (such as past fire suppression).[6]

Q **How can I learn about the expected local and regional consequences where I live?**

Scientists use sophisticated modeling to try to predict local and regional consequences around the world. Generally speaking, you can find some of these predictions simply by searching on the name of your region and climate change (for example, "Colorado and climate change"). For the United States, a particularly good source of information is the *Fourth National Climate Assessment, Volume 2: Impacts, Risks, and Adaptation in the United States* (online at nca2018.globalchange.gov), which provides detailed predictions about climate change for all regions of the United States. (A fifth U.S. national climate assessment should be available by the time you are reading this book.)

Storms and Extreme Weather

A second major consequence of global warming is more extreme weather events. As we've discussed, global warming really means an increase in energy in the atmosphere and oceans, and energy is what drives weather. With more energy, we expect hurricanes, thunderstorms, heat waves, and other extreme weather events to become more numerous, more severe, or both. Notice that extreme events include severe winter weather, leading to the ironic result that global warming can sometimes lead to cold spells and heavy winter snowfalls.

---

6 A good starting reference for learning more about attribution science is this article from the Columbia Climate School: news.climate.columbia.edu/2021/10/04/attribution-science-linking-climate-change-to -extreme-weather/

Q Can we tie particular storms to global warming?

Storms have always occurred, so we cannot claim that any particular storm was caused by global warming (though attribution studies can be used to investigate the role that global warming may have played). However, we can tie an overall trend to global warming. Many climate scientists use loaded dice as an analogy. Just as loading dice makes certain outcomes more likely than they would be with fair dice, global warming makes extreme weather events more likely than they would be otherwise.

Consider Hurricane Iota as an example (figure 3.8). This powerful storm (category 4) struck the Caribbean and Central America in mid-November 2020, when the hurricane season would typically have been over. Moreover, it was the seventh major hurricane of that season, which ended up as the most active Atlantic hurricane season on record, and it came just two weeks after another major hurricane had hit some of the same areas. We cannot say that global warming caused this particular storm or this particular hurricane season. What we *can* say is that global warming makes storms like this and seasons like this more likely, and that we can expect more years like this in the future. Cigarette smoking offers another analogy: We can't be sure that smoking caused a particular person's lung cancer, but we know that, on average, the more you smoke, the more likely you are to get lung cancer. In the

Figure 3.8. Hurricane Iota, viewed here from space on November 16, 2020, was the seventh major hurricane of the most active Atlantic hurricane season on record. Source: NOAA.

same way, as we add more greenhouse gases — and hence more energy — to the atmosphere, we should expect more extreme weather events.

Q Are heavy rain and snow events becoming more common?

You've probably noticed that news reports of major floods seem to be more common, and the data confirm an increase in heavy precipitation (rain or snow) events around the world. In fact, the increase in extreme downpours (and snowstorms) is occurring even in places that are becoming drier overall, such as the southwestern United States (figure 3.9). That is why climate scientists sometimes say that global warming is making it more likely that "when it rains, it pours" (or "when it snows, it blizzards"). The reason for this trend is easy to understand: Global warming means there is more total evaporation from the oceans, which means more moisture in the atmosphere and hence more rain (or snow) to fall.

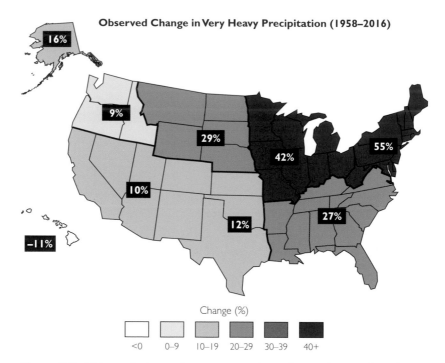

Figure 3.9. This graph shows the percentage change in very heavy rain and snow events for different regions of the United States from 1958 to 2016. Notice that the trend has been toward heavier events in all regions except Hawaii; in other words, when it rains, it pours (and when it snows, it blizzards).

Note: A "very heavy" precipitation event is defined here as an event that would have been in the 99th percentile for all nonzero precipitation days. Source: Fourth National Climate Assessment, Volume 1, Figure 7.4 (U.S. Global Change Research Program).

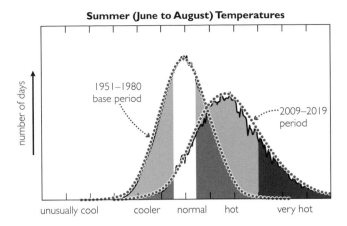

Summer (June to August) Temperatures

1951–1980
base period

2009–2019
period

number of days

unusually cool   cooler   normal   hot   very hot

Figure 3.10. These bell curves show the change that has occurred in typical summer temperatures for the period 2009–2019 compared to the baseline period 1951–1980; the labels along the bottom are for the baseline period. Notice that the entire curve has shifted significantly to the right (by approximately one full standard deviation). As a result, the recent "normal" summer day (the peak of the right-hand curve) is what would have been considered "hot" just a few decades ago.

Note: More technically, each curve represents temperatures normalized over all land locations in the Northern Hemisphere for the months of June through August; details can be found at columbia.edu/~jeh1/mailings/2020/20200706_ShiftingBellCurvesUpdated.pdf. Source: Data from James Hansen and Makiko Sato, Climate Science, Awareness and Solutions, Columbia University Earth Institute.

Are heat waves becoming more common?

Heat waves tend not to make as much news as major storms, but they can be deadly (particularly in places where air conditioning is rare). Heat waves therefore count as another type of extreme event, and the data show that they are indeed becoming more common. In fact, the increase in heat waves is part of a more general trend shown in figure 3.10. Notice that today's "normal" summer day is what was considered "hot" just a few decades ago, and "very hot" days are now quite common. The hotter days are exacerbated by the fact that nights have warmed even more, since this means we don't get as much of a chance to cool down at night as we did in the past. (The corresponding winter shift is even more dramatic, but somewhat less noticeable since winters are cooler.)

Do data confirm that extreme weather events are becoming more frequent?

It's not easy to define "extreme," and changes in human conditions can make storms that would once have been benign (because, for example, they affected unpopulated areas) seem more severe simply because there are now more people in their path. But even with

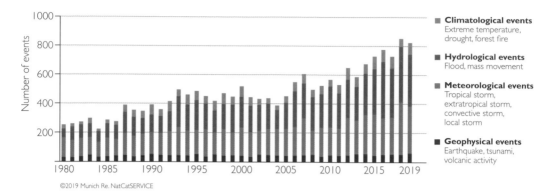

World Natural Catastrophes, 1980–2019

Figure 3.11. This graph shows a clear upward trend in the annual numbers of natural catastrophes (defined as those that involve significant property damage or loss of life) globally from 1980 through 2019. Except for the geophysical events represented by the small red portions of the bars at the bottom, all of the rest are weather- or climate-related.

Source: Met Office UK, based on data from Munich Reinsurance Company.

this ambiguity, the data strongly indicate an upward trend (figure 3.11), providing further evidence global warming is making extreme weather more common.

Q **What should I say if someone claims that a cold spell or snowstorm undermines the case for global warming?**

Remind them that while global warming tends to mean warmer winters on average, the general trend toward more extreme weather is expected to also include extreme winter storms. In particular, as we've already discussed, the fact that there is more moisture in the atmosphere leads directly to the idea that it is becoming more likely that "when it snows, it blizzards." Moreover, it's important to remember that local or regional weather events are not generally representative of global or long-term averages. Consider one famous case in which a United States Senator (James Inhofe) brought a snowball to Congress and used it to support his claim that global warming is a hoax. This occurred in February 2015, when there was a rare major snowstorm in Washington, D.C. — but the year as a whole ended up as the hottest on record to that point (see figure 1.15). Similarly, many parts of the United States experienced record-breaking cold in early 2023, but Europe was experiencing a record-setting winter heat wave at the same time.

Q **Do scientists understand how global warming can lead to winter cold spells?**

The details of the processes involved are complex, but the basic outlines are becoming clear. The driving mechanism is likely traceable to the fact that the Arctic is the fastest-warming region of the world (see figure 3.2), which means there is less temperature difference between the Arctic and lower latitudes than there was in the past. The accompanying reduction in summer ice coverage may also play a role. Although the details are still subject to debate, these changes appear to be weakening the high-altitude winds of the polar jet stream, allowing the jet stream to occasionally veer farther to the south. This, in turn, allows the very cold air of the "polar vortex" (essentially a cold air mass that generally circles around the north polar region) to flow southward as well, creating severe cold snaps.[7]

Melting of Sea Ice

It's fairly obvious that heat causes ice to melt, and we should therefore expect global warming to contribute to a reduction in ice cover around the world. Broadly speaking, there are two categories of ice melt, each with a different set of consequences:

1. The melting of sea ice, like that of the Arctic Ocean
2. The melting of *glacial ice*, meaning ice that is on land — especially in Greenland and Antarctica — and can therefore melt into rivers or the ocean

We'll focus on the melting of sea ice as our third major consequence of global warming, saving the melting of glacial ice for the discussion of sea level that follows.

Q Why is the decline of sea ice likely to be detrimental?

The good news is that the melting of sea ice does not have any significant effect on sea level, because this ice was already floating. The bad news is that it has many other detrimental effects. Most famously, it hurts polar bears, which depend on the ice for their hunting, but it also poses at least three global threats:

1. Amplification of global warming: Ice reflects much more sunlight than water does, so a reduction in sea ice means that Earth absorbs more heat from the Sun. This creates a *reinforcing feedback* that accelerates the melting and makes all the other consequences of global warming even worse.
2. Changes in ocean salinity (the amount of salt in the water): Melting sea ice adds fresh water[8] that lowers the salinity of the surrounding ocean, which can in turn lead to changes in ocean currents and nutrient levels. These changes could have dramatic effects both on coastal climates and on the productivity of fisheries that are a critical part of the global food supply.

---

7 This CNN article provides an excellent overview of these ideas: www.cnn.com/2023/02/03/world/extreme-cold-arctic-polar-vortex-climate-intl/index.html.
8 Sea ice is generally fresh water (or at least much fresher than normal seawater), because the freezing process tends to push out any salt. Fresh water also tends to be more acidic than sea water, so melting ice also adds to ocean acidification.

3. Changes in weather patterns: The reduced ice coverage may be contributing (in combination with the general lessening of the temperature difference between the Arctic and other latitudes) to changes in global weather patterns. As briefly noted earlier, these changes have already been implicated in some of the "weird" winter weather of recent years.

Q **Why doesn't melting sea ice affect sea level?**

Because the weight of floating ice already contributes to sea level, and its weight does not change as it melts. You can easily prove this for yourself by grabbing a cup of ice water (in which the ice is all floating). Mark the water level at the start and when the ice has all melted, and you will see that it does not change.

Q **What evidence indicates that sea ice is declining in the Arctic Ocean?**

Figure 3.12 shows data for Arctic ice coverage in September (when the sea ice is near its minimum after summer melting), along with a map comparing the 2012 coverage (the lowest to date) to two representative earlier years. The overall trend is clearly downward, and projecting it into the future suggests that by mid-century we could see at least some years in which the Arctic Ocean becomes ice-free in late summer. Data for other months show similar declines, though for winter months the decline is observed primarily through reduced thickness (rather than reduced area) of the ice.

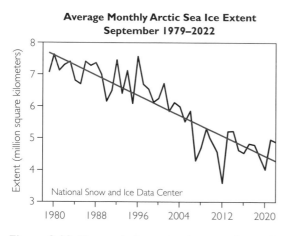

Figure 3.12. The graph shows the change in the total area of the Arctic covered by sea ice in September of each year since satellite records have been available; the black curve plots the actual data, and the blue line is a "best fit" line that shows the declining trend. Note that every one of the most recent 13 years (through 2022) has had lower ice coverage than any earlier year in the record. The map compares the overall boundaries of the September sea ice in three selected years: 1980 (red), 1998 (pink), and 2012 (white).

Source: National Snow and Ice Data Center; you can find data updated monthly at nsidc.org/arcticseaicenews/.

Sea Level Rise

A fourth major consequence of global warming is a rise in sea level. This rise comes from two distinct processes. First, water expands very slightly as it gets warmer, and while this *thermal expansion* isn't noticeable in a cup or bucket of water, the great depth of the oceans means it can cause a measurable rise in sea level. The second and potentially much greater contribution comes from melting of glacial ice, particularly in Greenland and Antarctica, which adds water to the oceans.

Q **How much has sea level already risen?**

Figure 3.13 shows measurements of the change in sea level since 1880, indicating an overall rise of more than 8 inches (20 centimeters). Estimates suggest that a little less than half of this increase has been due to thermal expansion, with the rest due to glacial ice melt.

Q Why do the satellite data and tidal data in figure 3.13 differ?

Toward the right side of figure 3.13, you can see that the tidal data show a slightly larger increase in sea level than the satellite data for the period through which both data sets are available. This is a result of the fact that the "sea level" we observe along a coastline can be affected by at least two different processes: (1) an increase in the level of the ocean water and (2) a change in the level of the coastal land relative to the ocean basin. Tidal gauges account for both processes, because they measure sea level relative to the land level along coastlines. In contrast, satellite data measure only the

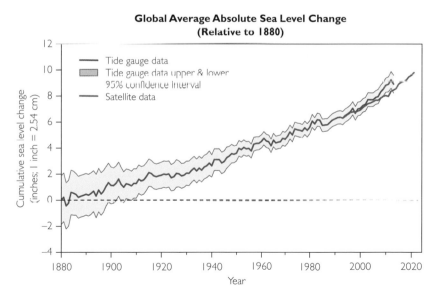

Figure 3.13. This graph shows measurements of the global average rise in sea level since 1880. The shaded region represents the uncertainty range for the data. Note the overall rise of more than 8 inches, or 20 centimeters.

Source: U.S. Environmental Protection Agency, based on data from NOAA and the Commonwealth Scientific and Industrial Research Organisation.

Figure 3.14. This photograph shows damage in New Jersey from the Hurricane Sandy storm surge (2012).
Source: Wikipedia/U.S. Air Force Master Sgt. Mark C. Olsen.

actual height of the ocean surface. The fact that the tidal data show a slightly greater average increase in sea level tells us that, on average, global coastlines have been sinking slightly.

A sea level rise of 8 inches (20 centimeters) may not sound like much, but it is significant for many low-lying regions. It's also important to note that the rise varies from place to place along the world's coastlines. For example, sea level along the eastern and gulf coasts of the United States has generally risen by more than the average, while sea level along the west coast and in Hawaii has risen by less than the average.[9] Moreover, the effects of even a small sea level rise can be magnified by tides and storms. Flooding at high tides is now estimated to be occurring at least two to three times as frequently as it did just a few decades ago, and storm surges are higher and go farther inland than they would otherwise, exacerbating the damage caused by tropical storms and hurricanes (figure 3.14).

Q Why does sea level vary in different places around the world?

Winds and ocean currents play a role, because they can essentially cause water to pile up and make sea level higher in some places and lower in others. In addition, land itself can rise or sink relative to the ocean basin. For example, the fact that coastlines are sinking on average (as noted earlier) is probably due to factors such as widespread withdrawal of groundwater and fossil fuel extraction, which remove mass that would otherwise keep the land level steady. Another factor is what is called "glacial rebound," in which land is still rising or falling as a result of the glacial retreat at the

9 You can find local rates of sea level rise at tidesandcurrents.noaa.gov/sltrends/sltrends.html.

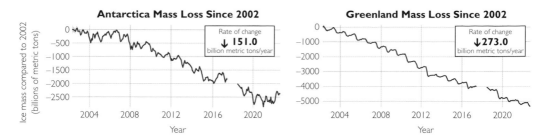

Figure 3.15. These graphs show measurements of the loss of ice from the Antarctic and Greenland ice sheets since 2002, as measured by NASA's GRACE satellites. The data values are negative because the total mass of ice is decreasing. Note: The data gap in 2017–2018 occurred because that was the time period between the loss of the first set of GRACE satellites and the launch of a second set.

Source: NASA; you can find the latest data at climate.nasa.gov/vital-signs/ice-sheets.

end of the last ice age. Consider North America, where land in the interior that was once covered by ice age glaciers is still rising up as it "rebounds" from the lifting of the weight with which the glaciers pressed down. In much the same way that puffing up the center of a pillow causes its edges to drop down, this rise of interior land is causing many coastal regions to sink down, which explains why the eastern and gulf coasts have seen a greater than average rise in sea level.

How do scientists measure the rate at which glaciers are melting?

Continued thermal expansion could easily add a few more inches of sea level rise in the future, but the total rise will depend primarily on glacial melting. It's therefore very important to have good measurements of how much melting is under way. For glaciers located in easily accessible places, such as those found in many mountain ranges, scientists can monitor the retreat directly (by, for example, comparing photographs taken at different times). But most of the world's glacial ice is located in Greenland and Antarctica, where such direct observations are much more difficult. Instead, since 2002, scientists have relied on an amazing set of measurements from NASA's GRACE mission.[10]

GRACE uses two satellites that together can measure very small changes in the local strength of gravity as they pass over different parts of Earth. Because changes in gravity arise from changes in mass, the GRACE measurements allow scientists to monitor changes in the masses of the ice sheets. Figure 3.15 shows GRACE data for ice melting in Greenland and Antarctica. Notice that substantial ice loss is happening in both places, but the rate of loss from Greenland is almost twice as large as the rate of loss from Antarctica, probably because temperatures

---

10 GRACE stands for the "Gravity Recovery And Climate Experiment." The original GRACE satellites operated from 2002 to 2017. A follow-on pair of satellites, known as GRACE-FO (for "Follow On"), began operations in 2018.

3

The Expected Consequences

have increased much more in the Arctic (see figure 3.2). (The GRACE satellite measurements can also be used to measure many other small changes in the distribution of mass on Earth, including changes that occur when flooding weighs down land after a storm.)

How much more can we expect sea level to rise in coming decades?

The future rise depends on two factors: (1) how much more carbon dioxide (and other greenhouse gases) we continue to add to the atmosphere and (2) exactly how the ice sheets respond to the resulting warming. The latter has proven to be notoriously difficult to understand, but recent research has helped scientists greatly reduce the uncertainties. Figure 3.16 shows what the latest models (as of 2022) project based on several different scenarios that combine the two factors. Notice the following key points:[11]

- Under the best-case ("low") scenario, sea level will still rise a bit further, but only by few inches. Of course, this can happen only if we rapidly reduce greenhouse gas emissions.

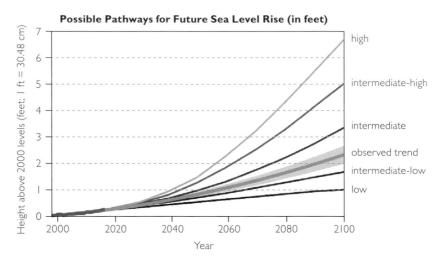

Figure 3.16. This graph shows actual sea level rise from 2000 to 2018, along with predictions for sea level rise through 2100 under different scenarios that consider a combination of future greenhouse emissions and assumptions about ice sheet response. The "observed trend" line simply extrapolates the recent trend in the observed data (see figure 3.13).

Notes: (1) The zero line here (for the year 2000) differs from that in figure 3.13 (which was based on 1880). (2) The "low" to "high" scenarios here are not the same as the ones shown later in figure 3.19 because these also include assumptions about ice sheet response. Source: NOAA Climate.gov, adapted from Sweet et al., 2022.

11 You can find a more detailed summary of recent findings on the impacts of sea level rise at oceanservice.noaa.gov/hazards/sealevelrise/sealevelrise-tech-report.html.

- If the trend of recent decades continues ("observed trend"), the sea level rise in just the next three decades will be about as large as that of the past century, and sea level will reach about 2 feet (60 cm) above 2000 levels by 2100.
- Under the worst-case ("high") scenario, sea level would rise by some 6½ feet (2 meters) by 2100 and would be on a pace to go far higher in the next century.

Given the impacts we are already seeing from sea level rise, anything beyond the "low" scenario is likely to have very damaging consequences for the hundreds of millions of people who live along the world's coastlines. As one example, figure 3.17 shows the regions of Florida that would be underwater with a 3-foot (1-meter) rise in sea level — which represents only the "intermediate" scenario from figure 3.16. The effects would be even more devastating in many other parts of the world.[12] Some island nations (and other inhabited islands) would likely end up completely underwater, and many of the nations that will be affected by sea level rise are much less equipped to deal with the consequences than a wealthy country like the United States. Many political scientists and military analysts fear that sea level change

Figure 3.17. The red regions on this map show portions of Florida that would be flooded by about a 3-foot (1-meter) rise in sea level.

Source: University of Arizona, Department of Geosciences. If you want to see how sea level rise would affect other regions, try the NOAA sea level rise viewer (coast.noaa.gov/slr/).

12 You can explore regional impacts of sea level rise with this interactive map tool: coast.noaa.gov/digital-coast/tools/slr.html.

alone could displace hundreds of millions of people from their homes, leading to political upheaval on top of climate upheaval.

Q Should I sell my beachfront real estate?

I'm not in the business of giving investment advice, and if you sell before sea level rises too much, you might make out quite well. But one thing is near certain: If your plan is to keep your beachfront property in your family for generations, your only hope of success lies in our rapidly stopping and then reversing the effects of global warming.

Q What if the polar caps melted completely?

If global warming continues unchecked, the polar caps might eventually melt completely, returning our planet to the ice-free state of dinosaur times. So if you really want to understand the long-term threat of sea level rise, you need to know how high sea level might go in this worst-case scenario. There are two general ways to answer this question. One is to look at geological data to see how much higher sea level has been in the past when Earth was ice-free. The other is simply to calculate the total volume of water in glacial ice and how much sea level would rise if that amount melted into the oceans. Both approaches yield the same dismaying answer: If the ice caps fully melt, sea level will rise by some 70 meters, or about 230 feet.

The good news is that this melting would almost certainly take thousands of years, which would in principle allow our descendants plenty of time to migrate inland as the coastline shifted. Still, if this melting ultimately occurs, it suggests the disconcerting possibility that future generations will have to send deep-sea divers to explore the underwater ruins of many of today's major cities.

Ocean Acidification

Our fifth and last major category of consequence is ocean acidification, which occurs as carbon dioxide dissolves in the oceans and, through a well-understood chain of chemical reactions, makes the oceans more acidic. Ocean acidification does not get much media attention but, in combination with the general ocean warming, it has been implicated in damage to coral reefs and disruption of ocean ecosystems. For example, greater acidity impedes the formation of new corals and makes it more difficult for oysters, clams, crabs, and other shellfish to form their shells. These changes, combined with pollution and overfishing, are likely to disrupt the food chain critical to sustaining the global fish stocks on which billions of the world's people rely for food and livelihood.

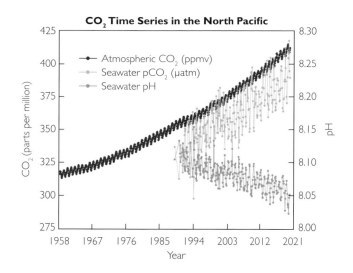

Figure 3.18. This graph shows evidence for ocean acidification. The red curve is the atmospheric carbon dioxide data (as in figure 1.8). The green curve plots measurements of the carbon dioxide concentration in the ocean. The blue curve represents the ocean pH, which is a measure of the water's acidity; lower pH means greater acidity. Notice that the pH goes down as the carbon dioxide concentration goes up, demonstrating that the ocean is becoming more acidic.

Source: NOAA; you can find the latest data at www.pmel.noaa.gov/co2/file/Hawaii+Carbon+Dioxide+Time-Series.

I won't say much more about ocean acidification, largely because the details require discussions of ocean chemistry and ecosystems that are beyond the level of this book. Nevertheless, it's important to keep in mind that because the oceans make up about three-fourths of Earth's surface, anything that affects the oceans is likely to affect the rest of the world as well. It is possible that ocean acidification could be as devastating in its effects on our civilization as any of our other four consequences, and it is therefore very important to consider it as part of any overall discussion of global warming.

What evidence indicates that the oceans are acidifying?
Figure 3.18 shows direct measurements of ocean acidification, leaving no doubt that it is real. The effects on coral reefs and other ocean ecosystems have also been observed and measured.

Future Climate Projections

We can expect all of the consequences we've discussed to become more severe for as long as the carbon dioxide concentration and global average temperature continue to rise. This fact leads us directly to key questions about what we might face in the future.

Q How much more warming will occur?

Scientists use models to predict a range of possible futures based on differing assumptions about future carbon dioxide emissions. Figure 3.19 shows three possible future pathways for these emissions, along with what recent models predict for the corresponding increases in global average temperature. Notice that only the "low" pathway would keep the total warming below about 2°C (3.6°F), and this pathway requires an almost immediate and dramatic reduction in carbon dioxide emissions. Other pathways would lead to greater warming and more damaging consequences.

Q How important is it to keep the warming below a threshold such as 1.5°C or 2°C?

These thresholds are often cited as critical levels of warming that we should aim to avoid. It's certainly true that holding the warming below these levels would mean fewer consequences than allowing these levels to be exceeded. That said, keep in mind that these levels are just convenient numbers for discussion, and what really matters is keeping the total warming as low as possible.

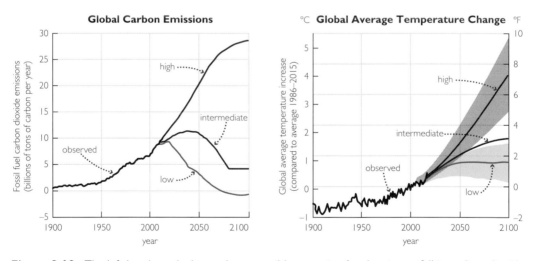

Figure 3.19. The left-hand graph shows three possible scenarios for the rise or fall in carbon dioxide emissions through 2100, and the right-hand graph indicates what models predict for future temperatures under each of those scenarios. On each graph, the black curve represents actual observations to date and the color curves represent low, intermediate, and high assumptions for future emissions.

Source: IPCC Climate Change 2021; the low, intermediate, and high curves shown here represent, respectively, IPCC model pathways SSP1-2.6, SSP2-4.5, and SSP5-8.5.

What is the danger of hitting a "tipping point" that could make things much worse?

Beyond all the threats that we have already discussed, many scientists are concerned about the possibility that we might reach some climate "tipping point" that could lead to even more devastating consequences. Examples of such tipping points include the following.[13]

- Arctic permafrost amplification: *Permafrost* refers to ground that remains frozen all year round and as a result contains large amounts of plant matter (from past growth seasons) that has not decayed. If ongoing warming causes this permafrost to thaw, the plant matter may then decay rapidly, releasing carbon dioxide (and methane) into the atmosphere and thereby accelerating the rise in the greenhouse gas concentration.

- Shifting currents: Some researchers worry that the fresh water entering the ocean from Arctic ice melting could alter major ocean currents, which would have significant effects on the global climate. For example, some evidence indicates that the Gulf Stream is already weakening, and a more severe breakdown of its flow would have drastic consequences for the climates of western Europe and the eastern United States and Canada.[14]

- Cascading ecological change: If regional climate changes happen fast enough, they might cause cascading ecological changes, such as losses of forests and biodiversity, that might in turn have severe consequences for the well-being of human populations.

- Rapid sea level rise: Some recent evidence suggests that a few vast glaciers in Greenland and Antarctica are undergoing melting in ways that could lead them to break apart and cause fairly rapid increases in sea level. For example, Antarctica's Thwaites glacier shows signs of potential disintegration, which could by itself cause sea level to rise by between 2 to 10 feet (0.6 to 3 meters) over a time span as short as just a few decades.

None of these or other potential tipping points are considered to be particularly likely, but neither can they be ruled out. The latest IPCC report puts it this way (in its summary point C.3):

_____

13   The following article has an excellent summary of these and other potential tipping points: grist.org /climate-tipping-points-amazon-greenland-boreal-forest/.

14   Shortly before this book went to press, a new study suggested that a breakdown of the ocean current system that includes the Gulf Stream (called the Atlantic Meridional Overturning Circulation, or AMOC) might occur by mid-century or even earlier. Reference: Ditlevsen, P. and Ditlevsen, S., *Nat Commun* 14, 4254, 2023, doi.org/10.1038/s41467-023-39810-w

Low-likelihood outcomes, such as ice sheet collapse, abrupt ocean circulation changes, some compound extreme events and warming substantially larger than the assessed very likely range of future warming cannot be ruled out and are part of risk assessment.

In other words, the level of threat we face from tipping points is difficult to assess. In the end, all we can really say is that the longer we let global warming continue, the greater the risk that we'll reach a tipping point.

How great is the overall threat posed by global warming?

Global warming is often said to be an "existential threat," which literally means a "threat to existence." But to the existence of what? It's important to be realistic about this and neither exaggerate nor underplay the threat.

On the exaggeration side, you may have heard some people claim that global warming poses a threat to life on Earth, but it doesn't. Even if we were to cause a sixth mass extinction (see page 74), life as a whole would surely survive, just as it did after other mass extinctions. It's also highly unlikely that global warming could drive humanity to extinction.

A more realistic statement is that *global warming poses an existential threat to our current prosperity.* Despite our many problems, the world today has achieved a level of prosperity far beyond that of our ancestors. We have made dramatic improvements in reducing famine and disease, child mortality is at an all-time low, and life expectancies are at an all-time high (not counting the temporary effects of the Covid pandemic). But all of this has been made possible in large part by a hospitable climate (and abundant energy), which global warming puts at risk. Note that this risk comes not only from the direct climate consequences that we've discussed, but also through the possibility of cascading effects. Climate change may make it increasingly difficult for us to feed a population that now exceeds 8 billion people, and we have already seen geopolitical instability arise when climate change leads to conflict or to the migration of "climate refugees"; for example, many military experts attributed the Syrian war and the rise of the terrorist Islamic State (ISIS) at least in part to a long-term drought that was almost certainly exacerbated by climate change. Indeed, the U.S. Department of Defense calls global warming a "threat multiplier."[15] That is why I think it is fair to say that global warming threatens our

---

15 See history.defense.gov/Portals/70/Documents/quadrennial/QDR2014.pdf.

current prosperity (and our hopes of building it further for those who don't yet fully enjoy it), and that at the extreme it could lead to events that might plunge our civilization into something reminiscent of past dark ages. While this may seem tame compared to the exaggerated threats of extinction or to all life on Earth, I doubt it's something that any of us would like to leave for future generations.

What's the bottom line for the danger posed by ongoing warming?

We are already experiencing significant consequences from our ongoing greenhouse gas emissions and the resulting global warming, and these consequences will continue to become more severe for as long as the warming continues. The bottom line therefore comes back to the question of "how much risk are you willing to take?" If we act now, it is still within our power to slow, stop, and ultimately reverse global warming. But the longer we wait, the more difficult this will be to achieve, and the greater the risk that we'll pass some irreversible tipping point. In my opinion, this makes it imperative that we act now, which is why we'll next turn our attention to the solutions that could allow us to build a true post–global warming future.

3

The Expected Consequences

4 Present and Future Solutions

The time for seeking global solutions is running out. We can find suitable solutions only if we act together and in agreement. There is therefore a clear, definitive and urgent ethical imperative to act.

— **Pope Francis, 2014 (message to the 20th Conference of the Parties, or COP 20)**[1]

Most media reports about global warming tend to focus on its consequences, which can make it look as if our possible futures range only from bleak to bleaker. But it doesn't have to be that way, because it is not yet too late to change our course. We already have available solutions that could allow us to stop the ongoing warming in its tracks, and other solutions are in development that will be able to do much more, including restoring the climate to a more natural state. Even better, I believe that these solutions present "win-win" scenarios that will not only protect the world for our children and grandchildren but that can also lead to a stronger economy with energy that is cheaper, safer, cleaner, and more abundant than the energy we use today. I'll therefore devote this chapter to a discussion of the solutions — and what it may take to implement them — that can pave the pathway to a post–global warming future.

Before we begin, I'll point out that this chapter will move us away from the "pure science" focus of the previous chapters and into more subjective discussions of technology and economics. In other words, while I believe that everything I've said up to this point in the book is scientifically defensible, with strong evidence behind it, this chapter will necessarily reveal some of my personal biases. I'll try to be clear about which statements are evidence-based and which reflect my opinions, and I hope that this will help you to make up your own mind about how we can meet Pope Francis's (above) imperative to "act together and in agreement."

---

1 The full text of the Pope's message can be found at vatican.va/content/francesco/en/messages/pont
-messages/2014/documents/papa-francesco_20141127_messaggio-lima-cop20.html.

Existing Energy Solutions

We humans need a lot of energy to maintain our modern civilization. We will need even more in the future, particularly if we hope to raise the living standards of billions of people living in the developing world. The key issue for global warming, then, is whether we can find this energy without continuing to add to our greenhouse gas emissions. This is a challenging task, because most of these emissions come from fossil fuels and these fuels represent the vast majority of the world's current energy supply (figure 4.1).

Nevertheless, we can in principle meet this challenge, because we already have technologies that could fulfill both present and future energy needs without continued use of fossil fuels. Broadly speaking, these technologies fall into three major categories that we'll discuss in the pages to come:

- energy efficiency, which allows us to get the same energy benefits with less energy usage
- renewable energy, such as wind and solar, which does not release greenhouse gases
- nuclear energy, which is often controversial but is a potential solution to global warming because it does not release greenhouse gases

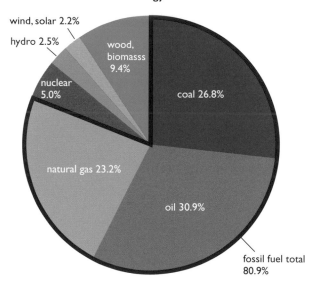

Figure 4.1. This pie chart shows the approximate makeup of the world's energy supply (as of 2020). Notice that the vast majority still comes from fossil fuels, illustrating the challenge of moving to alternative energy supplies.

Notes: "Hydro" is hydroelectric power, primarily from dams; "wood, biomass" includes wood, other crops, manure, and food waste that is burned for energy, primarily by people who do not have access to electricity. Data source: International Energy Agency.

Q Should we be talking about "energy" or "power"?

The two terms are often used interchangeably, but there is a technical difference between them. Power is the *rate* at which energy is used, released, or generated. You can understand the difference by thinking about a "100-watt" appliance. The 100 watts represents the appliance's *power* requirement whenever it is turned on, but the total *energy* it requires depends on how long it remains turned on. If you leave this appliance on for 10 hours, it will use twice as much energy as it uses in 5 hours, even though it draws the same 100 watts of power at all times.[2] I've tried to be careful to use the correct term where it matters. For example, you'll notice that I refer to *power* plants (such as a coal or nuclear power plant), because the rate at which the energy is generated matters in that case. In contrast, I use the term *energy* efficiency, because in this case we care about the total quantity of energy that is needed to obtain some benefit. But it often doesn't matter. For example, the pie chart in figure 4.1 would look the same if it showed global average power usage rather than energy usage, and it wouldn't really matter if I'd said renewable power rather than renewable energy (or nuclear power rather than nuclear energy) in the bullet list above.

Q How fast do we need to make the energy transition?

The answer depends on how much we want to limit future warming. For example, if we want to follow the "low" pathway in figure 3.19, which limits the total warming to about 2°C (3.6°F), we need to reach "net zero" emissions — meaning no further increases in the carbon dioxide concentration — before about 2070. I'll be more ambitious in our discussions and assume that our goal is net zero by 2050.

Q Couldn't we keep using fossil fuels if we implemented carbon capture technologies?

Proponents of this idea, often called *carbon capture and sequestration* (CCS), envision the use of technology that can capture the carbon dioxide released in the burning of fossil fuels and then permanently store (or "sequester") it, so that it never enters the atmosphere. (When applied to coal power, this idea is sometimes referred to as "clean coal.") The sequestration could in principle be accomplished either by injecting the captured carbon dioxide deep into the ground (in places where it would remain trapped) or by using it in chemical reactions that trap it in some non-gaseous form.[3]

The appeal of carbon capture and sequestration (particularly for those in the fossil fuel industry) is that, if it were successful, we could eliminate

---

2 This is why electricity bills usually show charges in units of "kilowatt-hours." 1 kilowatt is 1,000 watts, so 1 kilowatt-hour represents one hour's worth of energy drawn at a rate of 1,000 watts, or 10 hours' worth of energy drawn at a rate of 100 watts, and so on. For example, if electricity costs 15¢ per kilowatt-hour, it would cost you 15¢ to operate a 100-watt appliance for 10 hours. FYI, in case you want to convert to the standard international energy unit of joules: 1 kilowatt-hour = 3.6 million joules.

3 There are numerous possible ways in which this chemical entrapment might be accomplished, but the basic idea involves speeding up the processes through which most of the carbon dioxide that volcanoes have released through Earth's history became trapped in carbonate rocks. You'll recall from chapter 1 (see pages 14–15) that these processes explain why Earth has so much less carbon dioxide in its atmosphere than Venus.

most of our greenhouse gas emissions without needing to abandon the use of fossil fuels. However, I will not focus on this approach for three major reasons. First, while a few demonstration projects have shown that carbon capture and sequestration is possible (see, for example, the "NetPower" project), at present it could work only in places where geological storage of the captured carbon is readily accessible — which means it could not currently be applied to most existing emission sources. Second, even if someone eventually figures out how to apply it more broadly, it seems almost inevitable that capturing and sequestering emissions would cost more than allowing them to escape into the atmosphere. This added cost means that strong legal regulation and enforcement would be required to prevent unscrupulous operators from cheating and thereby undermining the goal of reducing emissions. While this type of regulation and enforcement is possible in principle, it would be difficult in the many places around the world that are still plagued by weak legal structures and corruption. Third, even if the carbon emissions were successfully captured, this would not automatically address other forms of air and water pollution associated with fossil fuels and their extraction. While these forms of pollution do not generally contribute to global warming, I personally think a broader goal of our energy transition should be to find sources that are as "clean" overall as possible.

That said, getting to net zero by 2050 is going to take an "all of the above" approach to limiting our greenhouse gas emissions, so I'm in favor of continued work to develop carbon capture and sequestration. This technology might prove cost effective in some cases, particularly if we implement significant carbon pricing (as we'll discuss starting on page 130), which would incentivize industries to eliminate their greenhouse gas emissions one way or another. It may prove especially useful for industrial processes that require very high temperatures, such as steel and cement production. Today, the high temperatures needed for these industrial processes are almost always generated by burning coal, and while some people are working on alternative ways of generating the necessary heat (such as with "green" hydrogen or small modular reactors, both of which we'll discuss later), many economists suspect that carbon capture and sequestration will prove to be more cost effective in this case.

Q What about using natural gas as a "bridge fuel"?

You've probably also heard the idea of switching from coal and oil to natural gas as a "bridge" between our current energy economy and a future, non-fossil energy economy. The major argument in favor of this idea is that natural gas releases *less* carbon dioxide than oil and coal (per unit of energy generated). In the United States, for example, carbon dioxide emissions have declined in recent years largely because many power plants have already been converted from coal to natural gas. But I see at least three general problems with this emphasis on natural gas. First, the costs associated with converting a power plant from coal to natural gas can take decades to recoup, making it more likely that natural gas becomes entrenched rather than a "bridge." Second, natural gas consists primarily of methane, which is a more potent greenhouse gas than carbon dioxide (see page 28), so leaks of methane during the process of natural gas production and transportation tend to offset at least some of the advantage that would otherwise be gained.[4] Third, a true solution to global warming requires not just reducing but actually stopping the release of carbon dioxide into the atmosphere, so the "less" carbon dioxide from natural gas is simply not good enough. For these reasons, I think it makes far more sense to focus on technologies that could completely end the release of greenhouse gases.

Energy Efficiency

The cheapest and easiest way to make headway against global warming is to reduce the demand for energy, which can be done by either (1) doing without some of the comforts we've become accustomed to or (2) improving the efficiency of our energy-using devices. I have friends and neighbors who have done remarkably well at the first strategy through such practices as walking or biking almost everywhere they go, hanging out clothes to dry instead of using a dryer, and turning their thermostats way down in winter. But while such dedication is admirable, it can be a tough sell to many other people. Indeed, while I like to think that I'm doing my part to help solve the problem, I still often drive even when I have other options, I use a clothes dryer and keep our home at a comfortable temperature, and I frequently travel by airplane. The realist in me therefore says that if we want to make big strides in reducing demand, we need to focus our attention on improving efficiency without demanding lifestyle sacrifices.

Can we achieve greater efficiency without lifestyle sacrifices?

Yes. To quote the definition given by physicist and environmental scientist Amory Lovins, energy efficiency means that we can "do the same or more with less." This is much more possible than you might guess, because there is a great deal of waste in our current energy usage. For example, incandescent light bulbs convert less than 5% of the energy they consume into light; the rest is wasted (primarily as heat). Similar waste is found in almost every other device we use, as well as in our electrical power grid[5] and in the fuel use of cars and airplanes. Therefore, if we can improve efficiency by reducing energy waste, we

---

4 Just to avoid any potential confusion: When natural gas is burned, methane combines with oxygen to form carbon dioxide. As a result, carbon dioxide is the only greenhouse gas released into the atmosphere by the *burning* of natural gas. The concern I'm addressing here is with methane that leaks directly into the atmosphere before it has a chance to be burned.
5 The "grid" basically refers to the entire system through which electricity is distributed. In other words, it is the network of power stations and power lines that moves electricity from the places where it is generated to the places where it is used.

can maintain (or improve) our lifestyles while significantly lowering our energy usage.

As an example, consider the energy used by buildings (both residential and commercial), which goes primarily into heating, cooling, lighting, and running other appliances. For heating and cooling, we can reduce demand simply by installing better insulation and windows; we can do even better by using more efficient technologies, such as heat pumps, for the heating and cooling that we still need. We can similarly reduce energy use by taking as much advantage as possible of natural light, by replacing old light bulbs with more efficient LED bulbs, and by upgrading to more efficient appliances. We can use smart technology to ensure that we use energy only when we actually need it. Note that besides saving energy, all these changes will also reduce your utility bills.

Overall, improved energy efficiency would seem to be a classic "no-brainer," because it saves both energy and money, with no reduction in benefits or lifestyle. Moreover, because greater efficiency means less total energy use, it can help reduce the future amount of global warming even if some of our energy is still coming from fossil fuels. And, of course, improved energy efficiency will make it easier to transition away from fossil fuels, since we'll need fewer new energy sources to replace them.

Q What are the "heat pumps" you mentioned?

Heat pumps work by transferring heat into your house in winter and out of your house in summer. Because they only move heat (rather than generating it), they are typically three times as efficient as more traditional heating and cooling systems; they are also much safer than using oil or gas, both in terms of accident risk and indoor air quality. The two most common types of heat pump are "air source" heat pumps, which transfer heat to or from the outside air, and "ground source" heat pumps (sometimes called "geothermal" heat pumps), which transfer heat to or from the ground. Ground source heat pumps generally require long water pipes run underground, which makes them easiest to install during construction. Air source heat pumps can usually be installed at almost any time, making them a great replacement heat source if your home is currently heated by electric baseboard or by gas, propane, or heating oil. Note that heat pumps are also now available for making your home hot water. Heat pumps can be expensive up front, but you may find tax credits or rebates that will lower their cost, and the fact that they are so much more efficient than traditional heating and cooling systems means that you'll save a lot of money over the long term.

You might be surprised (as I was) to learn that an air source heat pump can still be efficient at bringing heat into your house on a cold winter day (or dumping it out on a hot summer day), but that's the amazing part of the technology. Even very cold outside air still contains heat, and an air source heat pump can in essence collect and concentrate this heat until it is warm enough to heat your house and still be more efficient than traditional heating systems. Ground source heat pumps are generally even more efficient.

Q For transportation efficiency, how much does it help to switch to electric vehicles?

Any improvement in the fuel economy (gas mileage) of cars means lower gasoline usage and therefore lower carbon dioxide emissions. Replacing a gasoline car with an electric car is even better, for two reasons. First, electric car engines are much more efficient than gasoline-powered engines, so the same total amount of energy will go much further in an electric car, which generally means lower emissions even if the electricity for charging comes from a fossil fuel power plant. Second, if you get the electricity from a non-fossil source, such as solar panels on your roof, a wind power station, or a nuclear power station, then driving an electric car will not produce any greenhouse emissions at all.

There's one important caveat, however, related to what is sometimes called the "life cycle" energy cost of a car (or anything else), meaning the total energy usage "from cradle to grave." In the case of a car, this includes all the energy used in mining the materials for the car, assembling the car, driving the car, and ultimately disposing of or recycling the car components. These life cycle energy costs can be quite difficult to determine, since they depend on so many factors. For example, most mining today is powered by fossil fuels, which means that any new car (whether gas or electric) has already been responsible for a lot of fossil fuel usage even before you drive it, but the exact amount depends on where and how the mining was done. Nevertheless, most analyses that I've seen still conclude that electric cars reduce energy usage and emissions over the long term, and their impact will get even better if future mining can be done with energy from non-fossil sources.

Q Can you comment on the recent controversy over gas stoves?

The controversy concerns regulations that seek to limit the installation of gas stoves in new construction or to incentivize the replacement of existing gas stoves. (Contrary to some media reports, no serious legislation has been advanced to force the replacement of all existing gas stoves.) Many people have objected to these regulations, usually because they prefer gas cooking to the way that standard electrical cooktops work. Fair enough, but it's the wrong comparison, because you can now buy what are called *induction* cooktops. These are electric, but they work very differently from the old-fashioned electric cooktops in which you see a coil turn red as it heats up. Induction cooktops instead work by transferring energy through magnetic effects (which is why they work only with cookware that contains iron). This allows them to offer the same kind of precise and rapid heating control that people enjoy with gas stoves, plus they don't create waste heat (making them particularly attractive for commercial kitchens), don't release any indoor pollutants, and essentially eliminate fire risk. They can take a bit of getting used to, but overall, induction cooktops seem as good as or better than gas stoves in every way. The only significant drawback is that they are currently somewhat more expensive than gas stoves (though the price difference has been shrinking), but their greater energy efficiency generally makes them much cheaper in the long run, particularly if subsidies or tax incentives are available to help with the up-front cost.

Renewable Energy

Energy efficiency can take us a long way, but a full solution to global warming will still require replacements for fossil fuels, especially when we consider the growing demand for energy in developing nations. We'll start by considering renewable energy sources, such as wind, solar, geothermal, hydroelectric, and biofuels. They are called "renewable" because the energy comes from a source that can be used over and over again. While none of these are perfect — for example, toxic chemicals are used in solar panel production, wind turbines can kill birds and bats, and dams for hydroelectric power can damage river ecosystems — they have the advantage of producing energy without the release of greenhouse gases. The chief debates about renewables therefore focus on how much energy they can realistically provide and whether they are cost effective.

Q What is the potential of renewable energy?

The easiest way to think about the energy potential of renewables is by considering current total world power consumption, which is close to 20 terawatts.[6] The wind carries more than 10 times this much power around the globe, so in principle we could be fully powered by wind.[7]

The potential of solar is far greater: The total amount of solar power reaching Earth is almost 10,000 times current world power usage. In other words, if we could somehow tap only about 1/10,000 of the energy that reaches Earth in the form of sunlight, it could meet *all* of our energy needs.

---

6 You can follow this discussion without a full understanding of the units, but in case you are interested: A *terawatt* is 1 trillion watts, and 1 watt of power represents energy usage of 1 joule per second. So 20 terawatts represents 20 trillion joules of energy per second. Because there are about 8,760 hours in a year, this translates to a global annual energy usage of about 20 × 8,760 = 175,200 terawatt-hours (TWh). Note that if you look up global data, you'll also find many other energy units in use by different sources; to convert among them, try iea.org/data-and-statistics/data-tools/unit-converter.

7 However, there is considerable scientific debate about whether tapping this much wind power would have its own negative climate effects. These come from the fact that, in accord with the law of conservation of energy, tapping wind to move turbines reduces the kinetic energy of the winds. There's general agreement that we could safely tap a lot more wind power than we do at present, but this might not be the case if we tried to tap enough to meet the full current 20-terawatt global power demand. For more details, see esd.copernicus.org/articles/2/1/2011/esd-2-1-2011.pdf.

Q How much land would we have to cover with solar panels to tap this 1/10,000 of the incoming sunlight?

We'd need much more than 1/10,000 of the land because (1) the available solar energy varies with latitude, time of day, season, and local weather (cloudiness); (2) current solar panels are at best only about 20% efficient at converting sunlight into electricity; and (3) most of Earth's surface is ocean. Taking all these factors into account, estimates suggest that we'd need to cover roughly 1/250 of Earth's land surface with solar panels if that were our sole source of energy. This is a lot, but not inconceivable. For example, a recent estimate[8] found that we could supply roughly 15% of the world's total energy needs simply by putting solar panels on every available rooftop in the world.

Hydroelectric and geothermal power also offer significant potential. Hydroelectric already supplies about 2.5% of the world's total energy (see figure 4.1), though the downsides of building more dams suggest (to me, at least) that doing more with hydroelectric would probably require new technologies that could make use of tidal energy rather than rivers. For this reason, I won't say much more about hydroelectric, though tidal energy could become significant in the future.

Geothermal draws energy from Earth's own internal heat. Earth is quite hot inside (as becomes obvious when a volcano erupts), and in principle this heat could generate enough clean energy to meet all human energy needs for millions of years. The difficulty comes in tapping into this heat. At present, geothermal energy is easily accessible only near volcanoes, geysers, or hot springs. Iceland (a volcanic island), for example, now gets the vast majority of its energy from geothermal power plants. Obtaining geothermal energy at other locations requires deep drilling. Efforts are already under way to use the drilling processes pioneered by oil and gas companies for fracking to instead create so-called enhanced geothermal systems, which offer the potential to greatly expand geothermal energy.[9] Even better, if we could figure out how to drill much deeper into the Earth — to depths of about 20 kilometers (12 miles) — we could tap into geothermal heat almost anywhere on the planet. In that case, for example, we could drill down at sites of existing coal and gas power plants, making use of their existing steam turbines and electrical grid infrastructure but replacing their fossil fuel energy with carbon-free geothermal energy. We don't yet have the technology for such deep drilling, but some companies are working on it.[10] If they are successful, deep geothermal heat could

8 Joshi et al., 2021, doi.org/10.1038/s41467-021-25720-2
9 You can learn more about this technology with a search on "enhanced geothermal systems," but I'll point you to one company that is working on a project near where I live in Colorado: geothermal.tech.
10 See, for example, this article about a company called Quaise Energy: news.mit.edu/2022/quaise-energy -geothermal-0628.

be a game changer for our overall energy economy, not only because it would be so abundant but also because, unlike wind and solar, it would provide a constant (not intermittent) source of energy.

Overall, renewable sources clearly have the *potential* to meet all of our energy needs. The challenge lies in tapping this potential, particularly at a speed that could get us to net zero emissions by 2050. Because they seem to have the greatest potential (at least for the next couple decades), I'll focus our attention primarily on wind and solar, though geothermal (and possibly some other renewable technologies) may yet prove to be at least equally important.

Q What about ethanol and other biofuels?

Ethanol is one example of a biofuel, by which we mean a fuel made from crops, food waste, or other organic matter. A major reason for interest in biofuels is that they can often be used in existing engines (e.g., those of cars, buses, boats, and airplanes). Biofuels are renewable in the sense that we can always grow more of the crops (or other organic matter) used to make them. However, what we really care about with global warming is achieving net zero emissions. For biofuels, this means that growing the crops used to make them would need to absorb as much carbon dioxide from the atmosphere as is later released when the biofuels are burned. Unfortunately, most biofuels used to date do not meet this standard. In some cases the reason is that land has been cleared of forest to make room for the crops, and deforestation adds carbon dioxide to the atmosphere. In other cases the reason is that it can take a lot of energy to produce the biofuels, and this energy has generally come from fossil fuels. An additional drawback to using crops for biofuels is that they require land that is then not available for food production. For these reasons, I don't think that ethanol or most other existing biofuels have much potential for allowing us to achieve our net zero goals. However, as we'll discuss later, more advanced biofuels may be a different story.

Q Is it realistic to imagine getting all our energy from wind and solar?

Maybe, but the idea of achieving 100% renewable energy, particularly by our goal of about 2050, poses at least four major challenges. The first (and likely biggest) will be clear if you look back at figure 4.1, which shows that wind and solar represent only about 2% of the world's current power supply. We'd therefore have to increase their (combined) availability by a factor of close to 50 just to meet current world demand, let alone the higher global demand of the future. Still, there's at least some reason to think that we could do this. Figure 4.2 shows that global wind power capacity increased by a factor of more than 100 in the 25 years leading up to 2021, and similar data show that solar capacity increased even faster. Repeating this level of success in the next 25 years would therefore meet our goal, though continued growth at this exponential rate might be expected to become increasingly difficult.

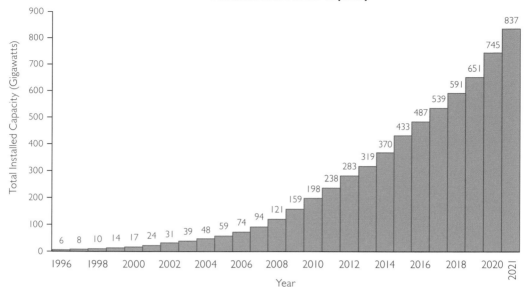

Global Installed Wind Capacity

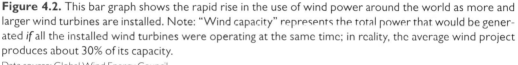

Figure 4.2. This bar graph shows the rapid rise in the use of wind power around the world as more and larger wind turbines are installed. Note: "Wind capacity" represents the total power that would be generated *if* all the installed wind turbines were operating at the same time; in reality, the average wind project produces about 30% of its capacity.

Data source: Global Wind Energy Council.

The second major challenge comes from the fact that wind and solar are intermittent (not steady) power sources, because wind turbines produce power only when the wind is blowing (and not blowing so hard that the turbines must be shut down for safety) and solar produces power only in daytime and when it's not cloudy. This is not a major problem when wind and solar are producing only a small fraction of the total electricity flowing through the grid, but scaling up wind and solar will almost certainly require a similarly large scaling up in the production of batteries (or other energy storage technologies) to store excess energy at some times and release it at other times. (Improving the power grid, so that it can distribute energy from areas that are generating energy to those that are not, could also help.)

The third challenge comes from the fact that wind and solar are generally used to produce electricity, which means they cannot directly replace the gasoline and similar fuels used for cars, trucks, ships, and airplanes or the gas or oil used to heat many homes and businesses. This is a major reason for the push to "electrify everything" so that we can power all our machines with electricity from the grid. Alternatively, we might use wind or solar power to produce a fuel such as hydrogen, which could more directly replace fossil fuels in some of those cases.

The fourth major challenge comes from the land use required to scale up wind and solar. While many communities have embraced large wind and solar installations (often as a source of extra revenue for farmers and ranchers, since these installations can often be installed with only limited impact on existing operations), other communities have mounted local resistance to these projects over issues such as aesthetics, land use, and noise. Regardless of whether you think these objections are valid, this type of resistance is likely to grow as more and more land is used for wind and solar projects.

Given these challenges, it's an open question as to whether 100% renewables is actually achievable. My personal guess is that it's probably not, which is why I think we need an "all of the above" approach to the problem of global warming. Still, data like that shown in figure 4.2 prove that we can make a lot of headway with wind and solar power, and the more we do with these and other renewables, the better off we'll be. Moreover, it's also possible that we'll be able to tap some game-changing renewable technology, such as the deep geothermal heat that I mentioned earlier (see page 103), hydrogen from the ground (see the second question below), or some other renewable source that has not yet gotten a lot of attention. If so, then 100% renewables might be more achievable than it currently seems.

Q You mentioned hydrogen as a potential fuel source; what could we do with it, and would it help with global warming?

Hydrogen is an excellent fuel because it is transportable and packs a lot of energy, which means it can in principle be used for almost any purpose for which we now use fossil fuels. Hydrogen can be used to produce power either by burning it directly or by using it in fuel cells (which use chemical reactions to generate electricity); either way, the only direct by-product is water.[11]

The challenge in terms of global warming comes from the fact that obtaining hydrogen in a useful form (as molecular hydrogen, or H_2) requires splitting water molecules — and this requires energy. If the energy for making the hydrogen comes from fossil fuels, as is the case with nearly all the hydrogen used as fuel today, then it doesn't help with global warming. However, if we use wind or solar power to split the water molecules, then the resulting hydrogen — sometimes called "green" hydrogen — can be used as a net zero fuel source. In that case, hydrogen could prove to be very important to our overall energy transition. Indeed, fuel cell cars already exist, though they have not caught on much because there is no existing infrastructure of hydrogen filling stations. (Electric cars have become much more popular, in part because they can be charged at home.) Many compa-

---

11 Water is the only by-product of fuel cells, making them extremely clean. Hydrogen combustion can lead to the production of nitrogen oxides, which are greenhouse gases, a result that is exacerbated if the hydrogen is burned in combination with natural gas. For this reason, fuel cells are generally the better option, unless the additional pollutants of combustion can be addressed. You can find details in this article: doi.org/10.1039/d1ea00037c.

nies are working to develop other uses for fuel cells, including in homes, cars, buses, trucks, airplanes, spacecraft, and more. Hydrogen produced with wind or solar power could also be used as the equivalent of a battery; for example, a solar farm could make hydrogen on sunny days and use it at night and on cloudy days, thereby smoothing out the energy flow.

Q **How would the potential of hydrogen change if we could get it out of the ground instead of having to make it by splitting water molecules?**

I'll be honest: Up until just a couple weeks before I added this question to the book, I would have said that this question was pointless, because there is no natural source of hydrogen. But then I came across an article in the prestigious journal *Science* (posted at doi.org/10.1126/science.adh1460) that caught me (and all of the colleagues whom I've since talked to about it) completely by surprise. Apparently, there *are* natural reservoirs of hydrogen underground, and the total amount of hydrogen is likely more than enough to meet all current and foreseeable global energy needs. Moreover, this naturally occurring hydrogen is just as clean as "green" hydrogen (because no energy is required to make it), and while scientists do not yet fully understand where it comes from, it is probably being continuously produced deep in Earth's crust, which means it is a renewable energy source. In other words, like deep geothermal heat, natural hydrogen from the ground is another potential game changer for renewable energy.

It is not yet clear whether it will be possible to drill for this hydrogen in the same way that we currently drill for oil and gas. But if it is, the necessary drilling technology will be essentially the same as that for oil and gas. So I'll add a little plea to oil and gas company executives: How about spending some big bucks to look into this option? After all, what could be better for your company than being able to reapply your existing technology to a new energy source that could also solve global warming?

Q **Are wind and solar cost effective?**

Yes. In the past, there were a lot of questions about whether wind and solar could compete economically with fossil fuels. Personally, I always felt this was a false debate, because as we'll discuss shortly, the true costs of fossil fuels are far higher than what we pay. But these questions have now become moot anyway, because the prices of wind and solar power have dropped so much that they are now generally cheaper than fossil fuels even without any government incentives or subsidies.

Q **What about the downsides of wind and solar, like turbines killing birds and bats and the mining for all the needed material?**

These are real downsides, but it's important to remember that no energy source is completely benign. We therefore have to consider energy sources by comparing their relative impacts, and renewables like wind and solar come out far ahead in any direct comparison with fossil fuels. Let's start with the mining. It's true that the mining needed for wind and solar power (and batteries that can store their energy) does a lot of ecological damage, but this mining in principle has to be done only once for each wind or solar installation (especially if you recycle the components when they eventually fail), while fossil fuels must be continually resupplied through ongoing coal mining or oil and gas extraction. As for the birds and bats, while wind turbines kill a lot of them, this number almost certainly pales in comparison to the number dying because of the climate change caused by our use of fossil fuels. And once you get beyond the mining and the birds and bats, it's no contest at all, since not only do wind and solar produce energy without greenhouse gas emissions, but they make essentially no pollutants at all and are therefore far safer and healthier on every level.

Nuclear Energy

Nuclear energy is our third existing potential solution to global warming. Because it generates a great deal of controversy, I'll go into a little more detail than usual to provide an overview of how it works and the concerns that surround it.

Q How does a nuclear power plant work?

The basic operation of today's nuclear power plants is pretty simple. They use a process called nuclear *fission* — in which a large atomic nucleus (most commonly of uranium) splits into two or more smaller ones — to generate heat, then use this heat to produce steam that drives turbines to generate electricity. The key to understanding the issues surrounding nuclear energy lies in details of the fission reactions.

Fission is one form of the naturally occurring process known as *radioactive decay*, which occurs because some elements and isotopes have nuclei that tend to break apart over time. We characterize the time it takes for a particular type of nucleus to break apart by what we call its *half-life*, which is the time it takes for half of any group of these nuclei to break apart. For example, the most common form (isotope) of uranium, called uranium-238, has a half-life of about 4.5 billion years; since that is also the approximate age of the Earth, we know that about half of the uranium-238 that existed on Earth when it formed is still present today.

All fission reactions release energy that can generate heat, but running a power plant requires much more heat than we can obtain from natural radioactive decay. Therefore, the primary goal of a nuclear power plant is to *induce* fission to occur at a far higher rate than it would occur naturally. This can be done only with a fairly small subset of all radioactive substances: namely, those that are capable of initiating a *chain reaction*, in which each individual fission releases two or more neutrons that can in principle cause additional fission reactions.

Although uranium-238 cannot do this, the much rarer uranium-235 can — but only if the atoms of uranium-235 are packed much more closely together than they are in nature, where uranium is generally found as a mix of more than 99% uranium-238 and less than 1% uranium-235.

Making fuel for a nuclear reactor therefore requires "enriching" the natural mix by separating out the uranium-235.[12] Enrichment is a complex and difficult process (generally involving large centrifuges), and the mix must typically be enriched to about 4% uranium-235 for use in nuclear reactors. Even then, the fuel mix won't sustain a chain reaction without extra help, such as a neutron source to start the reactions and a carefully designed mixture of other materials to help slow down the neutrons created during fission (because fast-moving neutrons will simply escape). These "extra" requirements for sustaining a chain reaction explain why building a nuclear reactor is a complex engineering project.

Q Is there a danger that the reactor fuel could explode as a nuclear bomb?

No. A nuclear bomb is in essence a chain reaction that goes out of control, and that is possible only when the nuclear fuel is far more concentrated than that used in nuclear reactors. For example, in the case of uranium, a nuclear bomb requires a mixture enriched to at least 90% uranium-235, so there is simply no way that the 4% enrichment mixture used in nuclear power plants could ever undergo the kind of sudden and uncontrolled chain reaction that makes a nuclear bomb. The absolute worst thing that can happen in a nuclear power plant is an accident leading to what is usually called a "meltdown," in which the nuclear fuel becomes hot enough to cause melting of some of the core material. This can lead to a release of dangerous radiation, and a steam explosion (as happened at Chernobyl) can release even more. But, again, there is no possible way for a nuclear power plant accident to cause a nuclear explosion.

Q Where does "radiation" come in?

Both natural radioactive decay and induced fission release high-energy particles and photons (such as x-rays or gamma rays) that we commonly refer to as *radiation*.[13] Note that, because uranium and other radioactive materials are found in natural rock, we are always exposed to at least some level

---

12 Nearly all current nuclear reactors use uranium-235 as their fuel source, but other possibilities include certain isotopes of plutonium and thorium. Plutonium is not found naturally on Earth, but it can be manufactured from uranium in "breeder" reactors. Thorium is actually more abundant and easier to obtain than uranium — and also has some safety advantages — so there is a lot of research currently going into the development of thorium-based nuclear reactors.

13 In physics, the term "radiation" refers to any form of energy carried by light or by subatomic particles, which means it includes harmless visible light and radio waves. Technically, then, we should distinguish harmless radiation from potentially hazardous radiation, but in the context of radioactive materials we will assume that we are talking about the hazardous type.

of natural radiation. We also get some radiation from space in the form of high-energy particles known as *cosmic rays*.

Q **How much natural radiation are we exposed to, and is it safe?**

Scientists use a unit called the sievert (Sv) to measure the overall biological effect of an absorbed dosage of radiation. Natural radiation exposure varies significantly around the world, depending on local geology and altitude. In most places, people receive a natural dosage of between about 1 and 10 millisieverts per year (mSv/yr), but there are some large cities where the annual dosage is above 100 mSv/yr. Importantly, careful studies have not found any significant health or life span differences based on these differing doses of natural radiation. In other words, this evidence suggests that doses up to 100 mSv/yr should be considered safe.[14]

Q **How much additional radiation would you be exposed to by working at or living near a nuclear power plant?**

Surprisingly little. While entering the core of a nuclear reactor (where the chain reactions occur) would be exceedingly dangerous, the structure that contains the reactor prevents this radiation from escaping. Data show that even if you work at a nuclear power plant, your annual radiation dosage will be less than 5 mSv/yr, which is well below the 100 mSv/yr that has been found safe in natural exposure. This explains why studies of people who have spent decades working at nuclear power plants have found no increases in cancer rates or decreases in life span. In terms of the surrounding community, the dosage becomes negligible. For example, measurements show that even if you lived next door to a nuclear power plant, your annual dosage would be only about 0.01 mSv/yr, which is far less than the dosage you receive from natural sources. To summarize, a properly operating nuclear power plant poses no radiation danger either to its workers or to its neighbors. The only exception would be in the event of a major accident, as we'll discuss shortly.

Q **What are the major pros of nuclear energy?**

Looking first at the positive side, there are four major reasons why nuclear energy has great potential as a replacement for fossil fuels, particularly when combined with energy efficiency and renewable energy:

1. Nuclear energy does not produce any greenhouse gases.[15]
2. Nuclear provides steady (not intermittent) power, which means it can plug directly into the existing grid without a need for batteries or other storage, and it can serve as a backup power source for intermittent sources like wind and solar.
3. Nuclear is by far the most "energy dense" of currently available sources. For example, a given weight of uranium used in a nuclear

14 Note, however, that if you live in an area with a relatively high amount of natural radon gas, this radioactive gas can become concentrated indoors, thereby raising the radiation dosage to dangerous levels. That is why radon mitigation systems are often needed in such areas.
15 The only greenhouse gas emissions associated with nuclear power come from the construction of the power plants and the mining and transportation of materials needed.

power plant produces nearly *3 million times* as much energy as the same weight of coal or oil. This means that comparatively little mining is needed to obtain nuclear fuel and that nuclear power plants have a much smaller overall "footprint" than other existing energy sources.

4. Nuclear is already the world's largest source of non-fossil energy (see figure 4.1), which means that scaling up nuclear energy to replace fossil fuels would be a much smaller challenge than scaling up renewable sources like wind and solar. Some countries already have very significant nuclear capabilities. France, for example, gets more than 70% of its electricity from nuclear power plants.

What are the major concerns with nuclear power, and how concerning are they?

Nuclear comes with some well-known concerns, including the danger of accidents, the problem of nuclear waste disposal, the security threat it may pose in terms of terrorism or nuclear weapons proliferation, and its economic viability. All of these concerns are valid, but they may not be as bad as commonly feared, especially when we consider nuclear energy in comparison to the fossil fuels from which we currently get most of our energy. So let's look briefly at each of these concerns, along with counterpoints to them.

> ***Concern — Nuclear Accidents:*** Nuclear power plants are safe when operating normally, but accidents can lead to the release of much more radiation. There have been three major accidents involving meltdown and significant radiation release: Three Mile Island (Pennsylvania, United States, 1979), Chernobyl (Ukraine, then part of the Soviet Union, 1986), and Fukushima (Japan, 2011). These have become etched in the public consciousness, and because there is no way to prevent all accidents, many people worry that nuclear power is simply too dangerous.

> ***Counterpoints:*** These accidents certainly should raise concern (though keep in mind that it is impossible for a nuclear power plant to explode like a nuclear bomb), but it is important to consider comparative dangers. While both Three Mile Island and Fukushima released radiation into the environment, there is no evidence that anyone received a dosage large enough to cause cancer or other shortening of life span.[16] The only known deaths

---

16 To be more specific, the U.S. Department of Energy found that the average radiation dose for people living near Three Mile Island was only about 1% of the natural background in the area and documented no health effects among workers. For Fukushima, both the World Health Organization and the United Nations Scientific Committee on the Effects of Atomic Radiation found no health effects that they could attribute to radiation, even among workers; however, the Japanese government cited one lung cancer death in a former worker as possibly having been due to radiation exposure.

from a nuclear power plant accident came from Chernobyl, where at least 30 workers died from radiation-related illnesses within months of the accident and it is estimated that the radiation has caused or will cause up to a few thousand more people from surrounding areas to die prematurely from radiation-induced cancers. That's obviously tragic, but it pales in comparison to the *millions* of premature deaths that occur every single year as a result of the air and water pollution caused by the use of fossil fuels.

In fact, per unit of energy generated, nuclear power appears to be about as safe as or safer than most renewable sources. For example, thousands of people have been killed by the failures of dams used for hydropower, and wind and solar can lead to deaths during the mining of their materials and from people falling from wind towers or rooftops. Figure 4.3 compares estimated death rates from various power sources.

Moreover, existing nuclear power plants have already been made safer as a result of lessons learned from the past accidents, and new designs have features that will vastly reduce the risk of accident in future power plants. In particular, nearly all existing nuclear power plants use what is called *active* cooling to prevent their nuclear fuel from overheating, which means that a failure of the cooling system can lead to a meltdown. Newer designs use

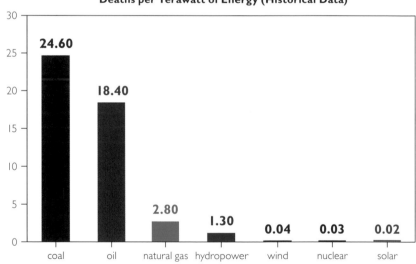

Figure 4.3. This bar chart shows death rates associated with various forms of energy production, based on historical data that take into account deaths from all causes (including accidents, pollution, and radiation).

Data source: Our World in Data, based on Markandya & Wilkinson (2007), Sovacool et al. (2016), UNSCEAR (2008, 2018), IPCC AR5 (2014), Pehl et al. (2017), and Ember Energy (2021).

passive technologies that will automatically shut down the reactor in the event of an accident or emergency, without the need for any human intervention.

Concern — Nuclear Waste: Just as you have to refill a car's fuel tank (or recharge its batteries), nuclear power plants need fuel "refills" as well, and that means the used (or "spent") fuel must be discarded. This discarded fuel is usually called *nuclear waste*, and some of this waste will remain dangerously radioactive for tens of thousands of years. It could therefore pose a danger to future generations unless we either find a way to keep it isolated enough that no one will ever come across it by accident or find a way to make it much less radioactive.

Counterpoints: There's no doubt that nuclear waste poses a challenge, but there are three main counterpoints. First, as with accidents, there's the comparative issue, and nuclear power again comes out far ahead of fossil fuels when amounts of waste are compared. In particular, the total amount of waste produced by a nuclear power plant in a year is what a typical coal-fired power plant creates *in a single hour* — and while the coal waste is much less radioactive, it is still very toxic. It's also worth noting that the high energy density of nuclear material means that the total amount of nuclear waste is quite small; a commonly cited statistic is that the total nuclear waste from all nuclear power generation in the United States to date could be stacked on a single football field to a depth less than the height of the goal posts. Indeed, it's possible that nuclear will even come out ahead of solar energy on the waste issue, because solar panels contain a variety of toxic metals and other toxic chemicals. Therefore, depending on the manufacturing process and whether and how the panels are ultimately recycled, we may end up with more total toxic waste from solar than from nuclear for equivalent amounts of energy generation.

Second, nuclear "waste" can in principle be used to generate further energy until the remaining radioactivity is extremely low. Indeed, a great deal of effort is currently going into development of "generation IV" nuclear reactors (sometimes also referred to as "fast" or "breeder" reactors) that could generate energy from today's nuclear waste without a need for more uranium mining. If widely implemented, this technology could almost completely eliminate the waste problem.

Finally, even if we were never to find a good solution to the disposal of nuclear waste, this waste will pose a danger only in the local regions where it is stored. While that is far from ideal, it still seems better than global warming, which affects the entire planet.

Concern — Terrorism or Weapons Proliferation: These concerns surround the possibility of sabotage, of theft of dangerously radioactive material, or of nuclear energy leading to the proliferation of nuclear weapons.

Counterpoints: These are real concerns, but they are also frequently misunderstood. From the standpoint of sabotage, remember that nuclear power plants cannot explode as nuclear bombs; the danger is sabotage that causes a meltdown and release of radiation. Prevention therefore requires great diligence in protecting nuclear power plants — but experience tells us that this is possible, since we successfully protect many other critical supplies and sites (such as airports and military installations). Moreover, a switch to passive cooling systems (as described above) could all but eliminate the sabotage threat. With regard to theft, unused nuclear fuel has a very low level of radioactivity, so a significant threat comes only from highly radioactive waste. Here again, it should be possible to protect the waste stream, and in the future, generation IV reactors (as discussed above) could leave any remaining waste so much less radioactive that it would no longer be dangerous.

With respect to nuclear proliferation, the primary risk comes from the fact that the same enrichment process used to make reactor-grade uranium could in principle be continued to the point where the uranium could be used for a bomb. However, the large machines used for enrichment are not easy to hide, so as long as any nation doing enrichment for nuclear power agrees to international inspections (which are currently conducted by the International Atomic Energy Agency), it should be possible to verify that enrichment is being done only to the point of making fuel for power plants, not for bombs. In the longer term, we might imagine an international agreement that allows enrichment only in a few places, so that it could be carefully monitored to ensure no bombs were built.

In fact, nuclear power could actually help the world reduce or eliminate nuclear weapons, because it is possible to dismantle nuclear weapons and then "de-enrich" their bomb material so that it can be used as reactor fuel. This has already been done: From 1991 to 2013, about 10% of the electricity used in the United States was generated by nuclear power with fuel recovered from dismantled nuclear bombs (as part of a project called

"Megatons to Megawatts," undertaken cooperatively between the United States and Russia). Estimates suggest that dismantling the world's remaining nuclear weapons would yield enough nuclear fuel to power much of the world for many years (and perhaps decades or centuries with generation IV reactors), which would be a fitting example of the Biblical call to beat "swords into ploughshares."

Concern — Economic Viability: The final commonly voiced concern about nuclear power is that it simply costs too much compared to other power sources, especially if you properly set aside money for dealing with the waste and the eventual decommissioning of the power plants.

Counterpoints: Cost has certainly slowed nuclear construction in the United States, though it's been less of an issue in many other nations, particularly in those that use a standardized nuclear design (as opposed to the case in the United States, where each nuclear power plant tends to have a fairly unique design). However, there are at least three reasons to think that nuclear might be much more cost competitive than is generally assumed. First, I believe that nuclear is already cheaper than fossil fuels when you take into account the many hidden costs ("externalities") of the latter (which we'll discuss shortly). Second, much of the high cost of nuclear power can be traced to public opposition (based on the concerns discussed above), which creates a risk of delay or cancellation of nuclear projects; if we educate the public about the misconceptions behind these concerns, this opposition might dissipate, allowing the cost to drop dramatically. Third, the nuclear industry seems poised for a technological revolution with what are often called small modular reactors (SMRs), which can in principle be mass produced and assembled to provide nuclear power at much lower cost almost anywhere and at almost any scale (figure 4.4). Numerous companies are working on SMRs, and though most are well behind their promised schedules, the chances of success still look quite good.

What's the bottom line for nuclear energy as a solution to global warming?

Many thoughtful people have considered all these issues and still come down on opposite sides of the question of whether we should build more nuclear power plants. I myself have vacillated over the years. However, I've ultimately come down on the side of "yes," for two major reasons. First, I'm unconvinced that we can build up renewables fast enough to reach our goal of net zero by 2050. We can do a

Figure 4.4. This illustration shows the size of a single module for a small modular reactor. Mass producing such modules may dramatically reduce the cost of nuclear energy.
Source: Oak Ridge National Laboratory, U.S. Dept. of Energy.

lot, but I doubt we can get to 100% renewable that quickly, and nuclear is the only other current option for filling the gap. Second, even if we can expand renewable energy much more rapidly than I'm expecting, it still has the problem of being intermittent (with the exception of geothermal), which means it generally needs "backup" from another power source. (This is likely true even with battery technology; for example, an extended cloudy period would drain solar batteries.) Today, that backup source is almost always fossil fuels, and nuclear is the only available alternative.

So my personal bottom line is this: At minimum, I think we should avoid closing any currently operating nuclear power plants, since that only makes it more difficult to reach our net zero goals. On top of that, I believe that we should be undertaking at least as much effort to increase our use of nuclear energy as we are to increase our use of wind and solar, because I believe that using efficiency, renewables, and nuclear in combination offers our best hope of rapidly achieving net zero.

Existing Technologies Summary

Given what we have discussed, it seems clear to me that we already have the technology necessary to end our fossil fuel dependence. Improved energy efficiency can take us a good deal of the way there, and some combination of renewables and nuclear can take us the rest of the way.

I won't claim that it would be easy, but it is possible, and in the process we'd not only eliminate most or all of the emissions that cause global warming, but also probably increase the energy supply, allowing us to maintain and improve living standards around the world.

How fast can the energy change happen?

I've been assuming a goal of net zero by 2050, which would require replacing fossil fuels with some combination of efficiency, renewables, and nuclear by that time. Is such rapid action realistic? As an answer, I'll relate a story about a time when I did my talk on this topic at a retirement home.

When we reached the point in the talk where I spoke about the need to change our energy sources, someone in the audience asked how fast this change could occur. I tossed the question back to the audience, and a 96-year-old former diplomat raised his hand and said "three to five years." I'd never heard someone claim that such a rapid transition was possible, so I asked him why he said that. His answer (paraphrased) was as follows: "Because I lived through World War II, and in order to win that war, we [the United States] managed to retool all of American industry — and to invent and build new industries — in only about three years. So if we treat global warming with the same urgency, three to five years should certainly be possible." I'll leave it at that, because I think his answer says it all.

What about the effects of greenhouse gases that are not from energy use, such as those from cement production, agriculture, and landfills?

Recall from chapter 1 that although carbon dioxide from fossil fuels is by far the largest source of human greenhouse gas emissions, there are also significant contributions of carbon dioxide from cement production and of methane and nitrous oxide from agriculture and landfills (see pages 27–28). A complete solution to global warming will therefore require that we also address these other sectors of our economy.

The good news is that solutions to these "other" problems are likely to be available. I won't go into much detail here, but I've already noted that new cement production processes may limit those emissions (see footnote 13, page 22), and I'll mention one other significant area of research: meat production, which is one of the major methane contributors in agriculture. You might at first guess that the only way to reduce these emissions would be by eating less meat, and that can certainly help. But it turns out that there are a number of ways to reduce emissions from livestock, and longer term it may be possible to eliminate meat-related emissions through so-called cultured (or lab-grown) meat, which is biologically identical to "real" meat but does not involve actual animals. Cultured meat could in principle be grown not only with no greenhouse gas emissions, but also with far less resource use than is required for meat from livestock; it would also eliminate the associated ethical issues and would likely even have health benefits, since live animals bring numerous disease risks.

The key point is that we can be optimistic that we'll soon find solutions to these "other" problems. In the meantime, because energy is the largest contributor to global warming, solving the energy issues eliminates most of the problem and therefore effectively buys us time to address the other issues.

Future Energy Solutions

Future energy technologies offer even greater promise. Some of these are "modest" extensions of existing technologies that are very likely to happen within the next couple of decades. For example:

- Advances in wind technology should make it possible to tap into generally steady offshore winds and perhaps also into high-altitude winds that are currently out of reach.
- Advances in solar photovoltaic technology should allow us to make solar panels that will convert a higher percentage of the sunlight they receive into electricity.
- Huge advancements are occurring in the technology for tidal power, which could make it a substantial source of energy in coming decades.
- Battery and other energy storage technologies are rapidly advancing, which will make it much easier to make use of intermittent sources such as wind and solar.
- Generation IV nuclear reactors are likely to become practical, which would in principle make it possible to vastly expand nuclear power while reducing the problem of nuclear waste and requiring little to no new uranium mining.

Even more exciting are a few technologies that are only beginning to be developed today. I've already mentioned potential game changers like deep geothermal energy and hydrogen from the ground. There are many more, and to give you a sense of the tremendous potential, I'll introduce you to three of my personal favorites: algae-based biofuels, solar energy from space, and nuclear fusion.

Q What are algae-based biofuels?

I noted earlier that ethanol and most other currently available biofuels have not proven to be carbon-neutral (at least as currently produced) and have the additional drawback of using land and water that could otherwise be used for food production. Future biofuels may be much better, especially those made from microbes such as algae. Algae are rich in natural oils that can be refined into substitutes for petroleum products, and they can be grown in a truly carbon-neutral way, in which they absorb as much carbon dioxide during growth as they later release when burned. In fact, it is even be possible to produce algae-based biofuels in a carbon-*negative* way, in which some of the carbon dioxide absorbed during growth is converted into a form (sometimes called "hydrochar") that prevents it from returning to the atmosphere. A further advantage of algae-based fuels is that they can be grown almost anywhere — perhaps even in floating, enclosed "ponds" in the ocean — so their cultivation should not have any negative consequences for the availability of food or fresh water. Moreover, because many forms of seaweed (e.g., kelp) are actually large collections of algae, it may be possible to obtain algae-based biofuels from expanded seaweed farming.

These attractive features have led to many efforts to produce these biofuels, but despite success at small scales, no one has yet succeeded in scaling up the production to a commercially viable level. Nevertheless, algae-based biofuels may yet prove to be an important energy source, especially as a carbon-neutral fuel source for existing airplanes (in which they have already been successfully tested), ships, rockets, and more.

What do you mean by solar energy from space?

Solar energy on Earth is intermittent because it works only in the daytime when it is not cloudy. But it's never cloudy in space, and if you put solar panels in a high enough orbit, it is never nighttime either. So another idea for solving our energy and global warming problems is to launch solar panels into high Earth orbit (figure 4.5), where they would absorb sunlight and beam the energy down to collecting stations on the ground. Given that the power of sunlight shining on Earth is nearly 10,000 times current world power usage, solar energy from space could easily meet almost any future energy need that we can imagine.

It would be expensive to launch solar panels into space, but perhaps not as expensive as you might guess. The "panels" for use in space might be no thicker than a thin film of plastic wrap, so enormous panels could potentially be unfurled from lightweight spools that could be launched by existing rockets. Moreover, these panels would not degrade in space the way they would on Earth, so once launched into orbit, they could provide energy for decades (or more) without the need for replacement. Overall, the total launch costs for enough panels to meet all global energy needs for decades to come might well be less than what we spend (globally) on energy in a single year at present.

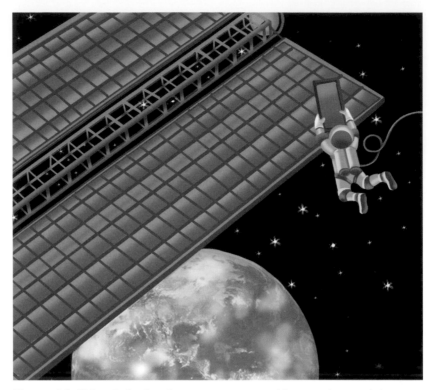

Figure 4.5. This painting imagines an astronaut working on solar panels in space, which are used to beam solar energy down to Earth.

Painting by Roberta Collier-Morales from *The Wizard Who Saved the World.*

The greater challenges probably lie in developing the technology for transmitting the energy to Earth, in building the collecting stations for that energy, and in tying those stations into a power grid that could distribute the energy around the world. But none of these challenges appear to be insurmountable. Several demonstration projects are already in development, including the U.S. Air Force's Space Solar Power Incremental Demonstrations and Research Project, the United Kingdom's Space Energy Initiative, and the European Space Agency's SOLARIS project; you can learn more about these simply by searching on their names.

Q What is fusion, and how does it offer truly incredible possibilities?

Recall that all current nuclear power plants are based on nuclear *fission*, in which atoms of heavy elements such as uranium are split apart. But what if we could instead tap nuclear *fusion*, which is the power source of the Sun and other stars?

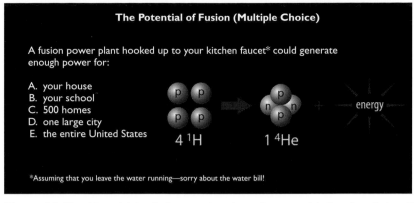

Figure 4.6. Try this multiple-choice question about the potential of nuclear fusion, if we can figure out how to tap it as a commercial energy source.

To get a sense of the tremendous energy potential of fusion, imagine for a moment that we had a miniature fusion reactor[17] that could do fusion in the same way as the Sun — by fusing ordinary hydrogen nuclei (which are individual protons) to make helium. We could supply the fusion reactor with fuel by extracting hydrogen from any source of water (since water is H_2O), so let's further imagine that you are willing to lend us your kitchen sink and let us use the water flow of your faucet. How much energy could we generate? Take a guess by trying the multiple-choice question in figure 4.6.

If you're like most students I've worked with, you'll probably take a guess from among choices A through D, since E sounds like an implausible throwaway. But E is the correct answer. Think about this fact: If we had the technological capability for fusion and you were willing to let us use your kitchen sink (and to leave the faucet water flowing), we could stop drilling for oil, stop digging for coal, dismantle all the dams on our rivers, take down all the wind turbines, and even turn off all the currently operating nuclear power plants. We'd be able to supply all the energy needed for the entire United States through the fusion of hydrogen extracted from the water flow of your kitchen faucet. There's a slight caveat: Rather than fusing ordinary hydrogen as the Sun does, future fusion reactors will probably start with the form (isotope) of hydrogen called *deuterium*,[18] which represents only about 1 in 6,400

---

17 The movie *Back to the Future* (1985) imagined a coffee-maker-sized device like this called "Mr. Fusion," though in reality it is unlikely that we could ever miniaturize fusion power plants in this way.
18 The reaction will probably also use the isotope of hydrogen called *tritium*, which will need to be manufactured because it is extremely rare in nature. There are several potential ways to do this, so it is not generally considered a significant obstacle to fusion power. It is also possible in principle to replace the tritium with helium-3, which is quite rare on Earth but could be obtained from the Moon.

hydrogen atoms. So you'd actually need about 6,400 kitchen faucets — or a small creek — to power the entire United States.

The tremendous power potential of fusion is only the beginning of its advantages. For example:

- Fusion not only is a carbon-free power source, but it produces no pollutants at all, since its end product is helium, which is safe, non-toxic, and useful.
- Because the fuel for fusion can be extracted from water, there is no need for mining.
- Fusion produces much less radioactive waste than fission,[19] and this waste decays more rapidly, so safely burying it away poses much less of a challenge.
- Fusion power is completely safe, because it is physically impossible for a nuclear fusion plant to go out of control and have a meltdown.
- Fusion power plants would not be prone to any of the other concerns we discussed for existing nuclear fission power either. For example, because the fuel and by-products are simply hydrogen and helium, they would be worthless to terrorists or bomb makers.

Given its incredible potential, you might wonder why we don't use fusion already. The answer is simple: Despite decades of effort, no one has yet figured out how to tap nuclear fusion for commercial power. But we *know* that fusion is possible (since it is the power source of the Sun and stars and is also used in thermonuclear bombs), so the development of commercially viable fusion comes down to a very difficult engineering challenge. Moreover, after decades of frustration, scientists and engineers working on fusion appear to be on the verge of great progress. For example, you've probably heard recent news reports of small breakthroughs in the path to fusion power, and a project called the International Thermonuclear Experimental Reactor (ITER) is currently under construction in France, with the goal of demonstrating commercially viable nuclear fusion within about a decade (figure 4.7). Numerous other research teams around the world, including many at universities and private companies, are also working to develop fusion power. There's no guarantee of success, but I'd be very surprised if we haven't met the fusion challenge by mid-century or soon after. And if we do that, we are looking at a future where fusion could provide us with a virtually unlimited source of inexpensive and ultra-clean energy.

---

19   The only radioactive waste is a by-product of neutrons released by the fusion reaction, and even this small amount of radioactive waste can be eliminated if the fusion reaction uses helium-3 (from the Moon) rather than tritium.

Figure 4.7. This photo shows construction under way inside the main building of the ITER project, which has a goal of demonstrating that fusion can be achieved in a commercially viable way.

Source: ITER Organization, iter.org.

Restoring the Climate Through Carbon Removal

Imagine that we successfully implement existing and future energy technologies to reach net zero by 2050 or sooner. That would obviously be great news, but we'd still be facing all the consequences of the high carbon dioxide level that we've already created in the atmosphere. Therefore, if we want to create a true post–global warming future, we will also need to find a way to bring the carbon dioxide level back down. *Keep in mind that, as we'll discuss below, we can get this type of full solution to global warming only if we combine carbon removal with net zero emissions.*

How low does the carbon dioxide concentration need to go (for a post–global warming future)?

Given that the carbon dioxide concentration was about 280 parts per million before humans started adding to it, you might at first think that this should be our target if we hope to restore the climate to a more natural state. However, most scientists don't think it is necessary to go quite that low. More specifically, renowned climate scientist James Hansen has used models to suggest that 350 parts per million would be an excellent target.

Q **Could we really restore the climate by bringing the carbon dioxide concentration back down?**

Yes, but with some caveats related to exactly what we mean by "restore." Some parts of the climate system would respond quite rapidly. In particular, we could expect the global average temperature to move back down as the carbon dioxide level falls, lagging the drop in carbon dioxide by no more than a decade or two. For example, if we could reach the target of 350 parts per million by 2080 or earlier, then by the end of this century the temperature could be restored to essentially what it was before we started messing with the climate. This should also mean that regional climates would generally be restored, and we could expect the number of extreme weather events to return to more natural levels. However, some aspects of the climate system would take longer to recover. For example, ocean acidification would probably take at least several decades to reverse (especially in deeper waters), and any sea level rise that had occurred would likely take centuries or millennia to recede. (For more details, see IPCC Climate Change 2021, The Physical Science Basis, FAQ 5.3.)

Q ## How might we remove carbon dioxide from the atmosphere?

The fact that today's carbon dioxide concentration is already above 420 parts per million (and rising rapidly) means that achieving a target like 350 will require not only stopping the increase by reaching net zero emissions but also removing carbon dioxide that is already in the atmosphere. Generally speaking, there are two basic ways to achieve carbon dioxide removal (CDR), which I'll refer to as passive and active.[20]

Passive removal uses natural processes to draw carbon dioxide from the atmosphere. The best known of the passive approaches is planting trees, which absorb carbon dioxide (through photosynthesis) as they grow. There are also ways to store more carbon dioxide in soil though agricultural methods known as "carbon farming" and a variety of land use policies. It may even be possible to get the oceans to take up more carbon dioxide from the atmosphere, perhaps through seaweed farming or, more controversially (because they might have their own detrimental consequences), through techniques such as adding iron to the oceans to spur phytoplankton growth or adding minerals that may help carbon dioxide be absorbed and stored. Some of these passive strategies seem quite promising, though they probably cannot by themselves bring the carbon dioxide level down as much as we would like.

This brings us to active removal strategies — technologies that could directly capture carbon dioxide from the air and then permanently

---

20 Another common way of categorizing carbon dioxide removal strategies is by whether they are biogenic (e.g., plants absorbing CO_2), geochemical (mineralization of CO_2), or synthetic (direct capture of CO_2). For a more technical overview of these strategies, see Küng et al., doi.org/10.26434/chemrxiv-2023-5f00r-v3.

store it away. We've already discussed how algae-based biofuels might be made in a carbon-negative way, and some scientists are working on genetically engineered microbes that might be even more effective at capturing and storing carbon dioxide. At least three other promising technologies might do far more. The first is *enhanced weathering*, which seeks to spur the absorption of carbon dioxide by minerals, in essence speeding up the natural processes that form carbonate rocks.[21] The second would stimulate electrochemical reactions in ocean water in order to turn dissolved carbon dioxide into mineral form, which would then cause the oceans to absorb more carbon dioxide from the atmosphere.[22] The third, which is already being used successfully in a demonstration project (figure 4.8), is *direct air capture and storage* (DACS), which uses machines to suck out carbon dioxide from air and then stores it either deep underground or by using chemical reactions to turn it into rock.

The primary challenge of active removal is the scale at which it would need to be implemented. For example, direct air capture and storage would require vast numbers of machines like those in figure 4.8, placed all around the world. Moreover, these machines require a lot of energy to run; by some estimates, running enough direct air cap-

Figure 4.8. This photo shows a set of modular machines in Iceland that are part of the world's first facility (named Orca) capable of direct air capture of carbon dioxide, with permanent storage deep underground. In principle, machines like this could help us bring Earth's atmospheric carbon dioxide level back down, at least if we first stop adding to it by achieving net zero emissions. Source: Climeworks.

21 See, for example, this article about a potential enhanced weathering project in Oman: scientificamerican.com/article/rare-mantle-rocks-in-oman-could-sequester-massive-amounts-of-co2/.
22 The UCLA SeaChange project has already developed a prototype of this idea; see youtu.be/WZuMRVjnkbU.

ture machines to make a significant dent in the carbon dioxide concentration might require up to half of the world's current total power usage, which means we'd need clean energy sources well beyond those available today. But you can probably see the good news here: If we act quickly to achieve net zero, buying us time to develop fusion or solar energy from space or other future energy technologies, then we should in fact have the necessary energy for active removal, and much more.

Q How much would we have to scale up active removal to create a measurable reduction in the atmospheric carbon dioxide concentration?

The answer obviously depends on the specific technology used, but let's assume direct air capture and storage like that of the Orca facility in figure 4.8. This facility is currently able to capture and store about 4,000 tons of carbon dioxide per year. Recall that 1 part per million (ppm) of carbon dioxide in the atmosphere represents about 8 billion tons (see page 25). Therefore, to use the same type of facility to reduce the carbon dioxide level by 1 ppm/yr would require scaling up the Orca capability by a factor of about 8 billion/4,000 = 2 million. This may sound daunting, but it's not impossible. For example, imagine that individual facilities could be 100 times the size of Orca, which is not unreasonable. We'd then need 20,000 of them around the world to accomplish the 1 ppm/yr reduction, which is fewer than the estimated 30,000 operating fossil fuel power plants around the world today. Bottom line: If we have a sufficient future energy source, such as fusion, and we are willing to build at the necessary scale, we are quite capable of bringing the carbon dioxide concentration down at a rate of 1 ppm per year or faster.

Q Couldn't we just wait for future carbon removal technologies rather than trying to stop the use of fossil fuels now?

Consider a medical analogy. Imagine that you had a potentially fatal disease for which scientists were working on a cure, but you knew that you could slow its progress if you quit smoking and improved your diet. Would you continue smoking and eating poorly in hopes that the cure would come before you died? I certainly hope not, and in the same way, it makes no sense to continue making the problem of global warming worse while we search for ways to reverse it — especially because any delay in reaching net zero increases the risk that we'll reach an irreversible tipping point. Moreover, the sooner we stop worsening the problem, the more likely it is that any future "cure" will be successful. So, as medical workers learn, first stop the bleeding.

Q What about geoengineering to cool the planet?

Geoengineering is generally defined as a large-scale intervention in the climate system designed to counteract global warming. There's a bit of semantic debate over whether the term should include the active carbon removal strategies we've discussed, but I will use it to mean schemes intended to cool the planet directly. The most common of these geoengineering schemes envision interventions such as seeding the atmosphere with aerosols that would reflect sunlight back to space or deploying giant sunshades in space, both of which would cool the planet by allowing less sunlight to reach the ground.

These ideas might at first sound reasonable, but they have at least three major drawbacks. First, because they do not automatically reduce our carbon dioxide emissions, they do not address the very serious problems of ocean acidification. Second, the proposed schemes require active maintenance; for example, the aerosol idea requires continually putting more aerosols in the atmosphere to replace those that rain out, and even the sunshades in space would likely need occasional orbital adjustments. If the maintenance ever failed — whether now or centuries from now — global warming would immediately resume, and if we'd continued adding carbon dioxide in the interim, it would be far worse than it is today. Third, geoengineering introduces global climate factors that do not exist naturally (such as side effects of the aerosols or a change in the sunlight distribution around the planet), making them difficult to account for in models. We therefore don't have any good way to predict the full consequences of these schemes, so even if they successfully stopped the rise in Earth's average temperature, we could not be confident that they wouldn't create other damaging forms of climate disruption.

Given these drawbacks, I would say that geoengineering is at best a "last resort" type of measure that should be considered only if we fail to make headway against global warming in other ways. It certainly should not be considered a desirable approach, and I hope you'll agree that it is far better to work toward a future of net zero emissions and carbon removal — which brings us to the topic of the economics that might spur this solution along.

Economics of an Energy Transition

I hope I've convinced you that solutions to the problems of global warming already exist and will become even better in the future. So you might wonder, why are we moving so slowly in implementing these solutions?

There are many factors involved, but my opinion is that the biggest impediment has been that prices for fossil fuels are artificially low, meaning that they do not reflect the true costs of these fuels to society. As a result, fossil fuels seem much cheaper than they really are, which tends to slow the adoption of other energy sources; in some cases, fossil fuel costs are so artificially low that they deter people from spending money on improved energy efficiency. The good news is that we are nevertheless seeing increasingly rapid progress in the transition to clean energy, largely because the prices for alternatives have been dropping rapidly. Still, I believe that the single most effective step we could take to speed action on global warming would be to make sure that prices for all energy sources reflect their true costs.

Q Why do you say that the true costs of fossil fuels are much higher than what we pay?

The answer is that current prices for fossil fuels do not include what economists call "externalities" but I'll refer to as *socialized costs*, because that's exactly what they are: real costs that are borne by society as a whole rather than by individual energy producers or users. Socialized costs are essentially equivalent to subsidies, because they reduce (subsidize) the prices paid by individual purchasers.

Consider the health costs of air pollution (figure 4.9), which is primarily a result of fossil fuels. Air pollution causes higher rates of numerous illnesses (such as asthma and heart and lung diseases), and various estimates hold it responsible for between about 4 and 7 million premature deaths annually around the world, including close to 100,000 in the United States.[23] These illnesses and deaths have real costs to society, including the related medical bills, the lost productivity of workers who are out sick or working at less than full capacity, and the statistically quantifiable costs for loss of life. None of those costs are built into the price of fossil fuels; instead, they are socialized through

Figure 4.9. These two photos were taken from the same location in Beijing, on a clear day and on a smoggy day. Poor air quality contributes to reduced life expectancy in many highly polluted cities.
Credit: Bobak Ha'Eri, Wikimedia Commons.

---

23 For example, a study published in the prestigious journal *PNAS* (Goodkind et al., 2019, doi.org/10.1073/pnas.1816102116) concluded that in 2011, small particulate pollution ($PM_{2.5}$) was responsible for 107,000 premature deaths in the United States, with a cost to society of $820 billion.

such things as taxes, medical insurance premiums, and reduced economic growth.

Other socialized costs of fossil fuels include the water pollution associated with their production and use, the environmental costs of strip mining for coal and fracking for oil and gas, the environmental and clean-up costs of oil spills, and the military costs of defending the global oil supply. There's also the fact that the United States and many other countries still provide many direct subsidies and tax write-offs to fossil fuel companies for their exploration and production of coal, oil, and natural gas.

On top of all that are two sets of socialized costs that are more difficult to quantify but may well be as great as or greater than all the others. The first is the intangible cost of the fact that nearly all major terrorist activities of recent decades have been funded largely with revenue from fossil fuels. The same idea also applies to many wars; in particular, Russia has used fossil fuel revenue to fund its invasion of Ukraine. Second, there are the costs from the consequences of global warming. For example, a storm that is more severe than it would have been without the warming has all the costs associated with additional property damage and loss of life, and many cities are already spending large sums to mitigate the effects of rising sea level.

What do the socialized costs add up to?

The uncertainties associated with the socialized costs (externalities) of fossil fuels mean there is a wide range of estimates for their total value, but even conservative economists have often come up with astonishingly high totals. In the United States, for example, an energy analyst who worked for both the Reagan administration and the conservative Heritage Foundation concluded in 2006 that the "hidden" costs of oil alone totaled $780 billion per year, and an earlier (2000) study by the U.S. Department of Energy concluded that dependence on imported oil had cost the United States about $7 trillion in lost wealth during the period from 1970 to 2000. More recent studies have estimated the heath costs of air pollution in the United States at hundreds of billions of dollars per year.[24]

How much would the "hidden" costs of oil add to the price of gasoline (in the United States)?

The answer obviously depends on the estimate you choose for the hidden costs, but for the sake of illustration let's assume the $780 billion cited

---

24 The first value cited in this paragraph came from the 2006 congressional testimony of Milton Copulos, available at govinfo.gov/content/pkg/CHRG-109shrg34739/html/CHRG-109shrg34739.htm. The DOE study can be found at osti.gov/biblio/763234. The pollution costs reflect many different studies I've examined, including the study from *PNAS* cited in the prior footnote.

above. (I'll use this value as is, but adjusted for inflation it would now be close to $1.2 trillion.) According to the U.S. Energy Information Administration, U.S. drivers used a total of about 135 billion gallons of gasoline in 2021. Therefore, the hidden costs would represent about $780 billion ÷ 135 billion gallons ≈ $5.80 per gallon. In other words, if we were paying the hidden costs at the gas pump, the price of gasoline would be nearly $6 per gallon *higher* than what we actually pay.

For global values, figure 4.10 shows estimates from the International Monetary Fund (IMF). Note the astonishing totals. In 2023, for example, the IMF estimates global fossil fuel subsidies at close to $7 trillion (with about $1 trillion of this attributable to the United States). The bottom line seems clear: *The true cost of fossil fuels is far higher than what we actually pay for them, very likely by at least a factor of two to three.*

Q What can we do about the market distortions created by the socialized costs?

The socialized costs distort the energy market by making fossil fuels seem much less expensive than they really are. Broadly speaking, there are two basic approaches to counteracting this market distortion. The first is to try to give a leg up to alternatives through government regulations (such as mandated levels of efficiency for cars and appliances)

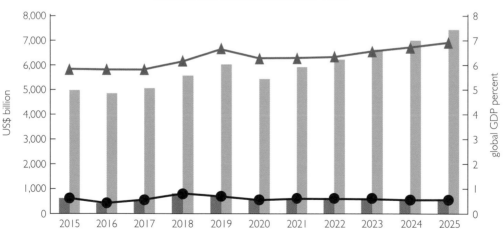

Figure 4.10. This graph shows the International Monetary Fund's estimates of annual total global subsidies for fossil fuels (values from 2021 on are projections). Although the match is not exact, the explicit subsidies on this graph generally represent direct subsidies and tax write-offs, and the implicit subsidies represent the socialized costs (externalities).

Note: The bars (which show values in dollars) and the triangles/circles (which represent percentage of global GDP) don't line up perfectly because global GDP changes over time. Source: International Monetary Fund, imf.org/en/Topics/climate-change/energy-subsidies.

and targeted subsidies or tax incentives (such as tax credits for installing solar panels or buying electric cars). The second is to incorporate some of the socialized costs into the actual price of fossil fuels through a carbon tax (or other form of carbon pricing) on all emissions of carbon dioxide and other greenhouse gases.

Interestingly, the first (regulatory) approach has been the one most widely used to date. For example, it is essentially the approach taken to dealing with climate change in the recent Inflation Reduction Act passed in the United States. However, economists across the political spectrum almost universally prefer the carbon tax approach, primarily because economic theory suggests that carbon pricing is the most efficient way to reduce emissions, meaning that it should allow us to reduce emissions further and at lower cost than we are likely to be able to through regulations. In addition, carbon pricing allows for a fairer comparison between the true prices of fossil fuels and other energy sources. Indeed, the benefits of a carbon tax are so widely accepted among economists of all political stripes that you'll rarely see any debate about whether it is the best approach. Instead, the debate focuses on two questions relating to the implementation of a carbon tax: (1) how high it should be and (2) what to do with the revenue it generates.

With regard to the first question, in principle the carbon tax should be high enough to account for *all* the socialized costs of fossil fuels. In practice, these costs are so high that, at least in the short term, we could not institute the appropriate tax without great risk to the economy. But this is an easy problem to deal with: We simply introduce the carbon tax gradually, so that individuals and companies have time to adapt as it rises.

The second question (what to do with the revenues) is where there is often more controversy, since taxes and revenues tend to be viewed quite differently by people with different political leanings. For this reason, I personally favor an idea known as "carbon fee and dividend," in which all the revenue is returned to the public through dividends distributed more or less equally to everyone. In other words, the government would not keep the revenue generated by the carbon tax (renamed a "fee" since the money isn't kept) but would instead use it to pay dividends. The big advantage of this approach is that it separates the question of appropriate spending levels from the question of how we solve the problem of global warming — which makes it far more likely to be acceptable to politicians of different political persuasions. It also offsets the otherwise regressive nature of a carbon tax because, at least on average, the less wealthy spend less on energy, which means their dividend checks will more than make up for the tax they pay. Of course, there will be exceptions (truck drivers, for example), so I hope our politicians would also implement policies to help such people, as well as to help those who lose jobs as a result of the transition away from fossil fuels.

Q Have any countries already implemented carbon pricing?

At least two dozen countries have some form of carbon pricing, either directly as a carbon tax or through a "cap and trade" approach (see below) that has similar overall effects. The World Bank produces an annual report (available at hdl.handle.net/10986/13334) in which you can see the details of existing carbon pricing programs.

Note: These plans usually show a price on emitted carbon, as opposed to carbon dioxide; see pages 24–25 to review how to convert between them. With that in mind, a good way to make sense of the carbon prices that you'll see in these plans is to remember that, roughly speaking, a tax of $10 per ton translates to about $0.10 per gallon of gas or about 1 cent per kilowatt-hour of coal power, and half that for natural gas power.

Q Where can I learn more about the "carbon fee and dividend" approach, and do you really think it's possible to get U.S. politicians to implement it in the current political environment?

Several groups have put forward detailed fee and dividend proposals and have websites that will answer almost any questions you might have. I particularly recommend the Citizens' Climate Lobby (citizensclimatelobby.org) and encourage anyone in the United States to get involved with this group. College students should also look at Students for Carbon Dividends (s4cd.org), and those looking for further proof that this approach can appeal across the political spectrum should check out the Republican-led Climate Leadership Council (clcouncil.org).

As to whether these plans could actually pass Congress, it would certainly be challenging, which is why it hasn't already happened. But I think that even for those who tend to oppose all new taxes, this case is different, both because all the revenue would be returned to the public and because the "fee" (carbon tax) is designed to correct for the market distortion created by the socialized costs of fossil fuels. If you think of it that way, our current system is in essence a "socialist" energy economy (since the socialized costs are so high), and most of the people who automatically oppose new taxes are also opposed to socialism. So if we focus on recouping the socialized costs, rather than on the "tax" (or fee) itself, perhaps we can get everyone to agree.[25] It's also worth noting that, because this market-based approach would make alternatives much more competitive with fossil fuels, we wouldn't really need regulations (and targeted subsidies and tax incentives) that are designed to do the same thing. So perhaps there could be a simple compromise between liberals and conservatives: Conservatives accept a carbon fee and dividend plan in exchange for liberals reducing or eliminating the regulations.

25 To this end, I'll offer an only partly facetious suggestion: Introduce the carbon fee and dividend bill with the name "SOS-energy" to stand for "Stop Overt Socialism in energy markets."

Q What about "cap and trade" instead of a carbon tax?

"Cap and trade" is an alternative approach that, like a carbon tax, is intended to encourage market forces that would lead to more investments in technologies that don't emit greenhouse gases. You can read more about this approach by searching on "carbon trading," but in brief it works like this: The government places a legal limit ("cap") on the total carbon dioxide emissions that are to be allowed, then sells, auctions, or initially gives away permits that allow companies to emit portions of this total. Companies can then buy and sell ("trade") these permits; for example, a company emitting more carbon dioxide than it is allowed could buy additional permits from a company emitting less than it is allowed. In principle, this encourages companies to reduce emissions, since they can profit by selling their permits to other companies that have been less innovative. If the cap is lowered over time, the total emissions should go down.

Cap and trade systems have been used successfully for other pollutants; a notable example is a system in the United States that has led to dramatic reductions in the emissions that cause acid rain. For greenhouse gas emissions, cap and trade has already been implemented in the European Union and several other nations and in the United States by California and the states involved in the Regional Greenhouse Gas Initiative. A bill to institute a national cap and trade system was passed by the U.S. House of Representatives in 2009 but died in the Senate.

Cap and trade can be a successful approach to reducing greenhouse gas emissions, and I don't see any reason to change it in those places that have already implemented this approach. However, my own opinion is that it is inferior to a straight carbon tax because (1) it is much more complex than a straight tax, since it requires the creation and maintenance of markets for the permits, and (2) unlike a carbon tax, which makes it easy for the public to see the true cost of the fuels they are purchasing, the behind-the-scenes trading of the cap and trade system leaves much of the public unsure why different energy sources cost what they do. So if you live in a country that has not yet instituted any carbon pricing, I'd push for the straight carbon tax (ideally with the "fee and dividend" approach). Of course, any type of carbon pricing is far better than nothing.

Q Doesn't a focus on carbon pricing for the U.S. (and other western countries) neglect the fact that China is now the world's largest emitter of greenhouse gases?

It's true that China is now the world's largest source of annual greenhouse gas emissions, though over history the U.S. is still by far the largest contributor to cumulative carbon emissions. Other developing nations are also rapidly catching up to the U.S. in annual emissions. This leads many to wonder whether carbon pricing in the U.S. would actually make much of a difference. I believe it would, for at least two reasons. First, the same arguments that explain why a carbon price would have overall economic benefits in the U.S. also apply to other countries. Therefore, by instituting a carbon price in the U.S., we would be demonstrating these many benefits by example, which would encourage other nations to implement similar pricing. Second, and again reflecting my personal belief in the power of the free market, I believe that carbon pricing would encourage such rapid innovation that the prices of clean energy sources would soon become much lower than those of fossil fuels, even without taking the carbon price into account. In that case, it is likely that other countries would adopt the same clean technologies, since the lower prices would make them an even more obvious choice for fulfilling future energy needs.

What's the bottom line for the economics of an energy transition?

To sum up our economic discussions, no matter how you look at it, the full costs of fossil fuels are not incorporated into their current market prices. Instead, these costs are socialized across society as a whole. In fact, they are even socialized across generations, because many real costs, especially those associated with global warming, will be borne primarily by our children and grandchildren (and subsequent generations).

The result is a distorted market for energy that has kept us addicted to fossil fuels even though we have available net zero alternatives (efficiency, renewables, nuclear) that would cost us less overall. Addressing the market distortion though carbon pricing would also encourage entrepreneurs and businesses to invest more in researching new technologies, because the potential payoff would be much more lucrative than it is when fossil fuel prices are artificially low.

That's why I've said that the energy transition away from fossil fuels is a "win-win." Making this transition will not only solve the problems associated with global warming and pollution but also improve the economy because we'll be getting the same (or more) energy at lower total cost. So let's stop accepting the huge risks of climate change, and instead start moving toward a future that will be better and brighter for everyone.

5

A Pathway to a Post– Global Warming Future

I want you to panic.
I want you to feel the fear I feel every day.
And then I want you to act.

– Greta Thunberg, age 16 (at the 2019 World Economic Forum)

In this short book, I have only touched on the many details that lie behind the issues of global warming, its consequences, and its potential solutions. Nevertheless, I hope that by now I have convinced you that, despite what you may sometimes hear in the media and from politicians who choose to remain ignorant, the basic facts are simple and clear. Global warming is under way, it is caused by the burning of fossil fuels and other human activities, and left unchecked it poses an existential threat to the prosperity to which we all aspire. That is why Greta Thunberg (quoted above), like so many other young people today, is filled with fear about our climate future.

But this brings us to the most important part of Greta's quote, which is the call to action — because, as we've discussed, the future is not just a choice between "bleak and bleaker." If we act wisely and rapidly, we have the opportunity to put the world on a pathway to a post–global warming future that can be every bit as amazing as any future that science fiction writers have ever imagined. And it requires only two "simple" steps:

1. Achieve net zero greenhouse emissions quickly, so that we stop making all the problems of global warming worse and alleviate the risk of hitting some tipping point.
2. Develop new energy sources (such as fusion or solar energy from space) and carbon dioxide removal technologies that will then allow us to lower the atmospheric carbon dioxide level, thereby restoring the climate to a more natural state.

What does a post–global warming future look like? Let's take the viewpoint of today's youth. If medical advances continue to increase life expectancy at the rate at which it increased in the last century, by 2100 the average person will live to around age 100. This means that anyone born since about the year 2000 — which includes everyone of college age and younger — can reasonably expect to still be alive in the year 2100 and beyond. Now imagine that, by mid-century, we've changed our energy economy to carbon-free sources, so that the carbon dioxide concentration is no longer rising. Further imagine that we've achieved fusion (or some other technology for clean and abundant energy) by this time and have begun active carbon dioxide removal. Then by 2100, the youth of today could be living in a world in which

- The natural climate has been largely restored, making global warming a topic for history books.
- We have used our abundant energy (and appropriate policies) to eradicate global poverty and to raise living standards for everyone.
- Fresh water availability is no longer a problem, because we can use our clean and abundant energy to desalinate seawater.
- We are also using the energy to power new agricultural practices (such as vertical farming and production of "cultured meat") that provide abundant, healthy food with far less land use and ecological damage.
- The abundant energy is powering human exploration of the solar system, and with the mineral resources available from the Moon and asteroids, we no longer need to do mining on Earth. This has allowed us to restore global forests and ecosystems and to turn much of our planet into a global set of "national parks" for people around the world to enjoy.
- And much, much more.

This incredible future really is open to today's youth — and for the rest of us, to our children and grandchildren. But it isn't a given, and the window of time in which it is possible is closing rapidly. We need to act now, and we need to act with urgency. With that in mind, I'll close by suggesting a pair of more personal ways of thinking about the issue, one for the younger crowd and one for the older.

For the younger folks, I suggest trying this exercise (figure 5.1):

Imagine it is your 100th birthday, and you are celebrating a life that has been everything you might hope for in a world in which the problem of global warming has been solved. Write a letter from your 100-year-old self to the actual you of today, giving yourself the encouragement you need to set yourself on a path to this great life. Be sure to tell yourself about some of the things you did to help make a better world, and about at least some of the amazing things you've seen happen in your long life.

Figure 5.1. Assuming advances in medical science continue, the average person will likely live to age 100 by the end of this century, which means that most of today's youth can expect to be alive in the year 2100 and beyond. So if you are in this age group, try the exercise shown here to help you think about your own role in creating a post–global warming future.

For those of us who are somewhat older, I'll suggest a slightly different exercise, one that is designed to help overcome the political divisiveness that has often accompanied this issue. This is possible because no matter where you fall from right to left on the political spectrum, there's one thing that I believe we can all agree on: We all want a better world for our children and grandchildren, and for everyone else who will follow us in the future. So consider this fact: On average, we are all about 50 years older than our grandchildren are or will be. (If you don't have or expect to have grandchildren, think instead about grandnieces/nephews or the grandchildren of friends.) For example, if you're reading this in 2025 and you are in your prime career years today, your grandchildren will be in theirs in 2075. If you are a senior citizen today, your grandchildren will be approaching their own senior citizenship in 2075. So imagine sealing a letter in a time capsule for your grandchildren to open 50 years from now, like the sample one in figure 5.2. When they open the letter, how will they feel about what you did — or did not do — in the face of today's understanding of the choices that will determine our climate future?

> — To be opened in 50 years —
>
> Dear Grandchildren,
>
> As I write this in 20XX, many people are arguing about whether global warming is a real problem and, if so, how serious it will be for you when you are my age. I have examined the evidence, and I have decided to . . . [fill in your decision]
>
>
>
> Hope your world is a good one.
>
> Love,
>
> _____
> [your name here]

Figure 5.2. On average, 50 years from now your grandchildren will be about as old as you are today. Complete this letter and put it in a time capsule for them to open in 50 years. Then ask yourself, how will they feel about what you did — or did not do — to help secure their future?

In case you are wondering, my own way of completing the letter is simple. I will continue to do everything I can to support climate education both in schools and among the public, with particular emphasis on the simple underlying science of global warming and what it will take to build a post–global warming future; I will do everything in my power to push for a carbon tax like the one discussed in chapter 4, since I believe that is the clearest and simplest way to incentivize the changes needed to build such a future; and I will support only those politicians who acknowledge the reality of global warming and are willing to take serious steps to protect the futures of our children and grandchildren. I hope you will join me in making similar commitments to the future.

Acknowledgments

Although I'm listed as the author of this book, I hardly deserve the credit, since all the research I describe has been done by others. So my first acknowledgment goes to the many scientists around the world who have dedicated their lives to helping us understand global warming in all its details and implications. I'm also deeply indebted to the many scientists who have put simple and clear information out on the web, especially those behind the websites listed on the Additional Resources pages that follow. These sites were invaluable to me in researching the many details included in this book.

The pedagogical approach of this book is one that I've developed with a great deal of help from numerous other people, especially my textbook coauthors Nick Schneider, Megan Donahue, and Mark Voit. Additional thanks to Dr. Schneider for coming up with the four "levels of denial" that form the backbone of chapter 2.

Both editions of this book have greatly benefited from careful review by many readers. I'll start with Steve Montzka (NOAA), who did the first expert review of the manuscript and gave me many great suggestions for improvements and then reviewed again for the second edition. Additional expert reviewers who provided great assistance on one or both editions include Scott Mandia, William Becker, David Bailey, David Bookbinder, James McKay, Kirsten Meymaris, Greg Meymaris, Shawn Beckman, Yoram Bauman, John Bergman, Minda Berbeco, Glenn Branch, Piers Forster, William Gail, Dave Cleary, Roger Briggs, Travis Rector, John Birks, David Gruen, Jenny Wilkinson, Tom Blees, Chris Packham, Bob Epstein, Dave Siskind, Callum Thompson, Jim Hooton, Ed Dlugokencky, and Rudy Kashar.

For producing the book, I thank Mark Ong and Susan Riley of Side By Side Studios in San Francisco. Mark and Susan not only did all the design work, but also provided many great suggestions on the approach and content. The book website (globalwarmingprimer.com) looks great thanks to web maestro Courtney Faust of Saffron Design in Boulder, Colorado. Finally, I thank my wife, Lisa, and my adult children, Grant and Brooke, for their ongoing support, inspiration, and insights.

Additional Resources

There are far too many great resources on global warming for me to attempt to list them all in one place, but I'll offer a few suggestions. First, I'll note two resources of my own for different audiences:

- **Children's Book:** I hope you'll check out my book *The Wizard Who Saved the World* (Big Kid Science), which explains the basic science of global warming while offering children an optimistic and inspirational view of our future and how they can contribute to it. Note: The book was selected for the Story Time From Space program, in which it was read from the International Space Station by Japanese astronaut Koichi Wakata (photo on next page). A video of his reading is posted at StoryTimeFromSpace.com.
- **Free Middle/High School Climate Curriculum:** I have created a free, online, digital textbook for Earth and Space Science — including a full chapter (chapter 7) on global warming science, consequences, and solutions — designed to meet all NGSS middle school science standards for this subject area; much of the material is also suitable for high school courses. This free resource includes everything teachers need to teach their courses, including activities, assessments, and extensive notes to help teachers prepare. Please check it out at grade8science.com.

Websites: The following are a few of my personal favorites among the many great global warming websites.

- ipcc.ch: This is the place to find the latest reports from the Intergovernmental Panel on Climate Change (IPCC), which represent the work of thousands of climate scientists from around the world.
- globalchange.gov: This is the home page for the U.S. Global Change Research Program, and it is where you will find the Fifth National Climate Assessment, which should be available by the time this book is in print.
- climate.nasa.gov: NASA maintains this site focused on global climate change with outstanding summaries and regularly updated news.

Japanese astronaut Koichi Wakata reads *The Wizard Who Saved the World*, a children's book that discusses the science of and solutions to global warming, aboard the International Space Station. A video of the reading is posted on the website of the Story Time From Space program (www.StoryTimeFromSpace.com).

- climate.gov: This is the NOAA website where you can find the latest data on almost any aspect of climate change.
- skepticalscience.com: For answers to questions beyond those addressed in this book, the Skeptical Science website is the place to go. Especially see their list of "climate myths" in the left column.
- desmog.com: This website is dedicated to combatting climate disinformation. Whenever I hear some questionable claim, I use this site as a quick way to learn about the credibility (or lack thereof) of the individuals or organizations behind the claim.
- theclocktowerproject.org: This cool project has produced an awesome full-dome planetarium movie that puts climate change in the context of the long scale of time.
- history.aip.org/climate: This website expands on Spencer Weart's fantastic book *The Discovery of Global Warming* (Harvard University Press), which helped me learn about the history behind this topic.
- astronomersforplanet.earth: "There is No Planet B" is the tag line for this group of astronomy professionals, educators, students, and amateurs that is focused on climate education and action.

In addition, here are a few more websites that I use frequently as references:

- carbonbrief.org
- realclimate.org

- climatecentral.org
- rmi.org
- yaleclimateconnections.org
- climatecommunication.org
- csas.earth.columbia.edu

Note: There are many more, but be careful, as there are also many disinformation websites that have legitimate-sounding names.

For those interested in looking beyond global warming to other important issues that will affect our children and grandchildren, I suggest checking out www.contractwiththefuture.org.

Finally, you can find additional materials, including information about booking me to speak on this topic, at the book website:

www.globalwarmingprimer.com

Index

About the Author

Human history is more and more a race between education and catastrophe.
— H. G. Wells, 1920

Dr. Jeffrey Bennett (B.A., Biophysics, University of California, San Diego; MS., Ph.D., Astrophysics, University of Colorado) has devoted his career to science and mathematics education and outreach. He has taught at every level from preschool through graduate school, including more than 50 college classes in astronomy, physics, mathematics, and education. He is the lead author of college textbooks in astronomy, astrobiology, mathematics, and statistics that together have sold more than 3 million copies; of critically acclaimed books for educators and the public on topics including Einstein's theory of relativity, the search for extraterrestrial life, and math and science teaching; and of seven award-winning science books for children, all of which have been selected for the Story Time From Space program, in which books are launched to the International Space Station and read aloud by astronauts. Other career highlights include serving two years as a visiting senior scientist at NASA headquarters, proposing and co-leading development of the Voyage Scale Model Solar System on the National Mall in Washington, D.C., creating the free Totality app for learning about solar eclipses, and creating a free online digital textbook for middle school Earth and Space Science. Among other awards, he is a recipient of the American Institute of Physics Science Communication Award and, most recently, the 2023 Klopsteg Memorial Award for education from the American Association of Physics Teachers. For more details, please see his personal website, jeffreybennett.com, or his primary outreach website, bigkidscience.com.